Building Math Minds:
Spatial Reasoning with Blocks

Created by Matthew Les

ISBN: 978-1-7380909-1-4

Math is Key

To access video tutorials demonstrating the solutions to these questions, please visit www.mathiskey.com/solutions

Alternatively, you can search for "Math is Key: Spatial Reasoning with Blocks" on YouTube to find the videos uploaded by the "Math is Key" channel.

To access resources regarding additional math topics, visit www.mathiskey.com

Contents

Adjacent Blocks

Count how many blocks are adjacent to the shaded block in each pile.

Two blocks are adjacent if the face of one block is touching the face of another block, even by a little bit. They are not adjacent if only their edges are touching. Every block in the pile is the same size and shape. The dimensions of the blocks are 4 inches by 4 inches by 1 inch. Some blocks are hidden from view. No block is floating completely in the air; if a block is not resting on the ground, there must be another block underneath holding it in place. For some of the questions, the top view will be provided showing a "bird's eye view" of the pile of blocks.

Example:

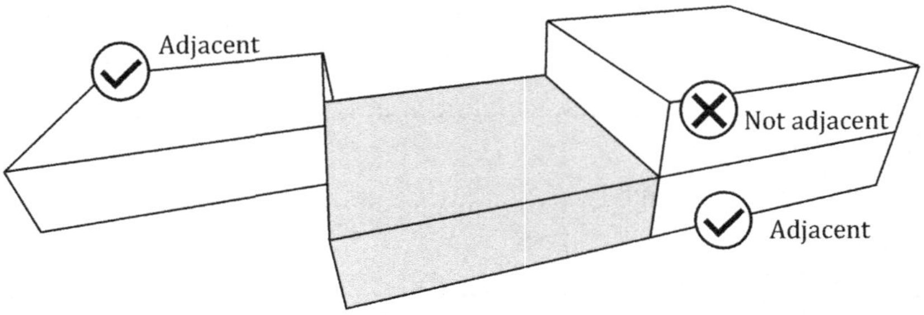

Count: 2

1.

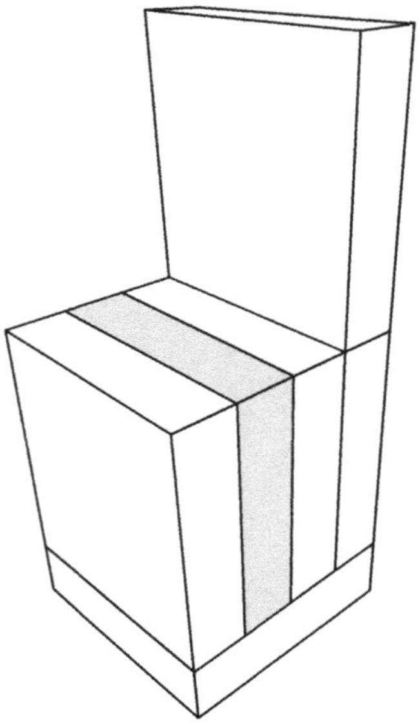

Count: ________________

2.

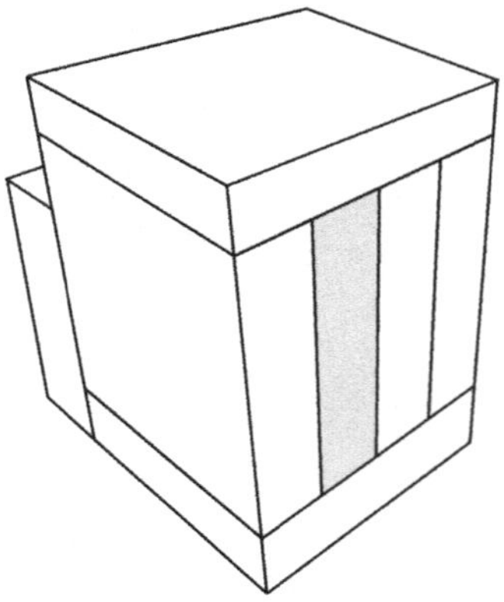

Count: ________________

3.

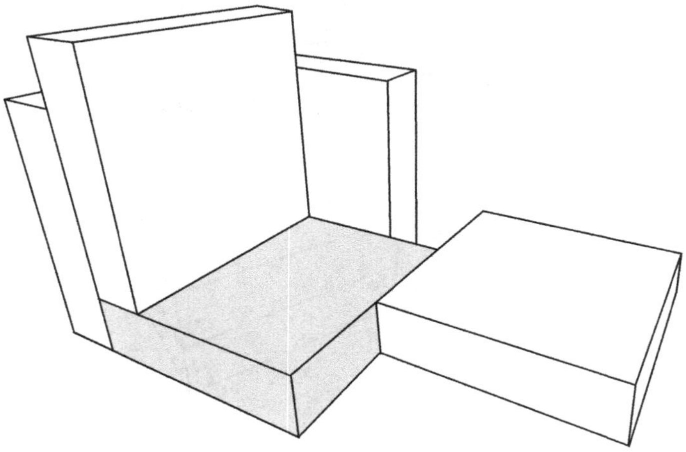

Count: ______________________

4.

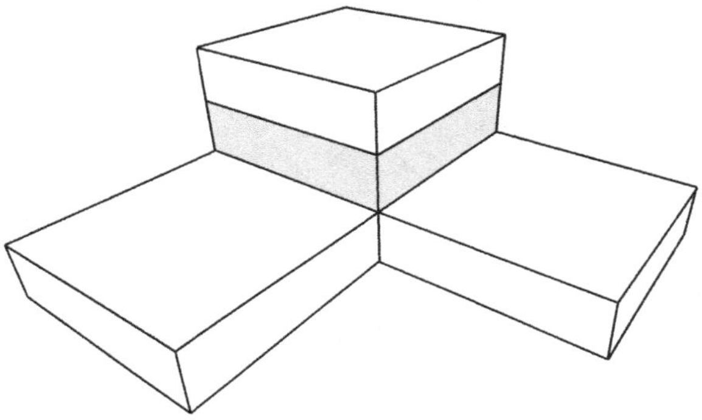

Count: ______________________

5.

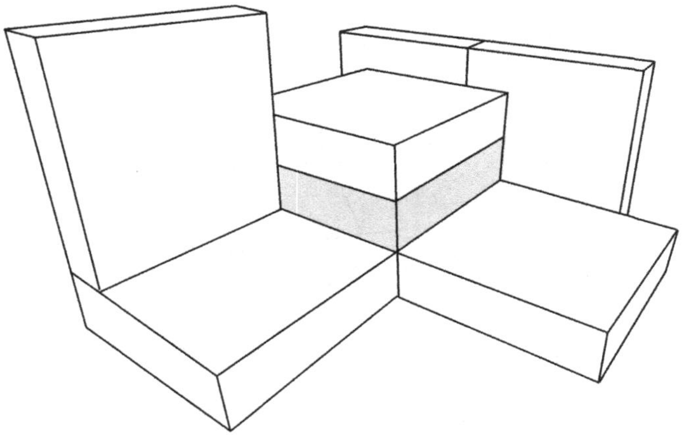

Count: ______________________

6.

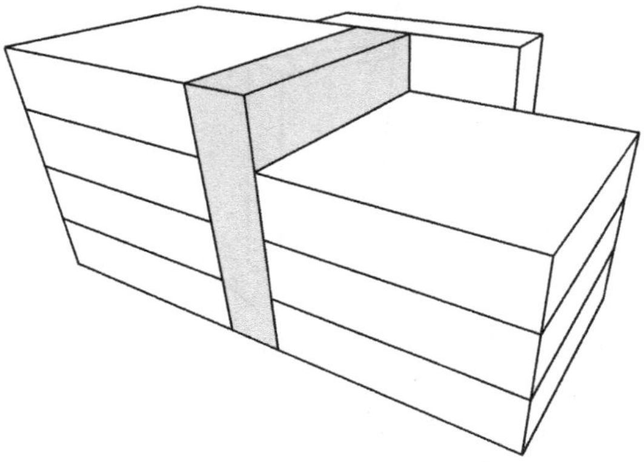

Count: ______________________

7.

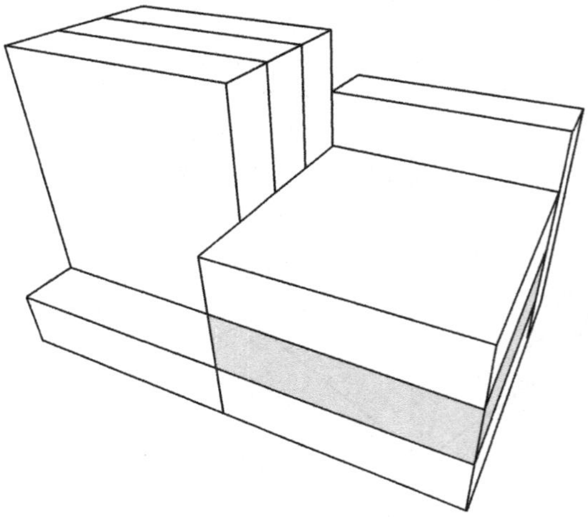

Count: ______________________

8.

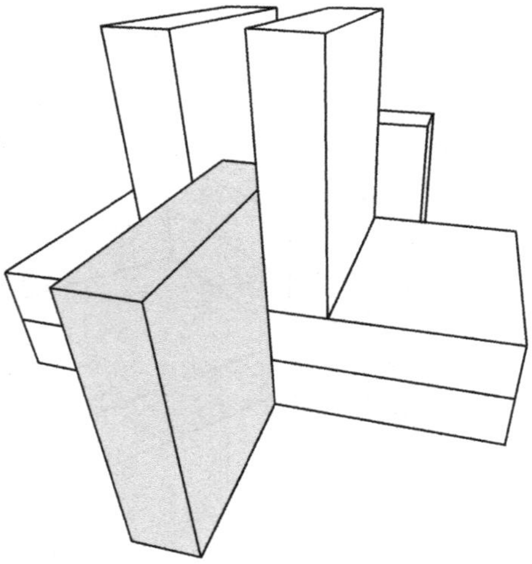

Count: ______________________

9.

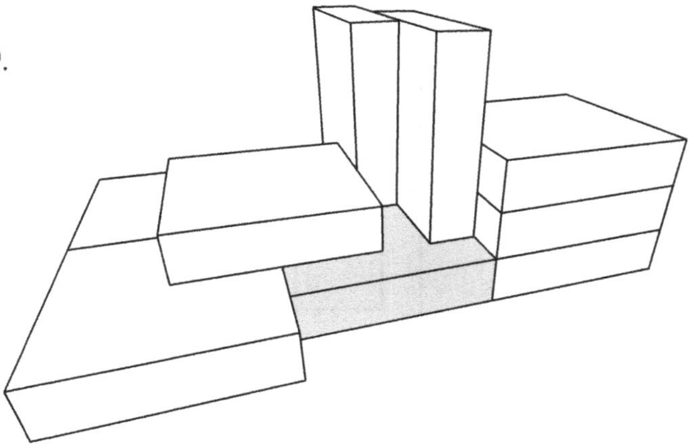

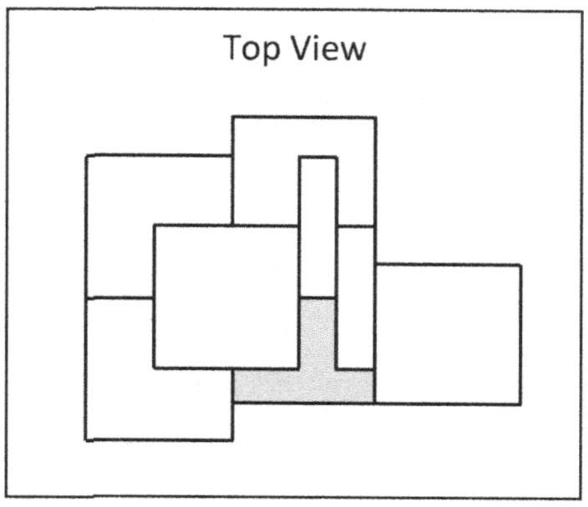

Count: ____________________

10.

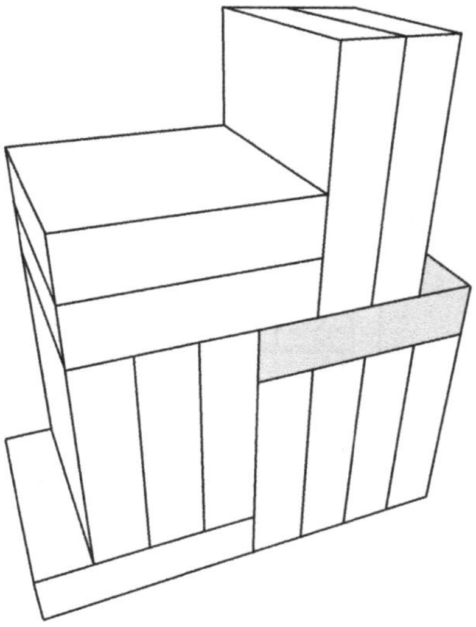

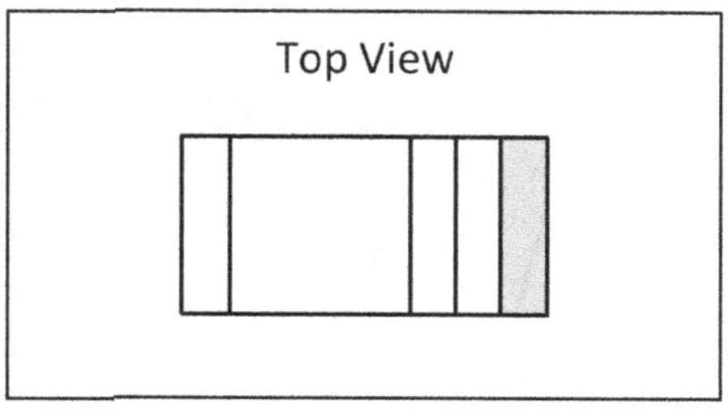

Count: ____________________

11.

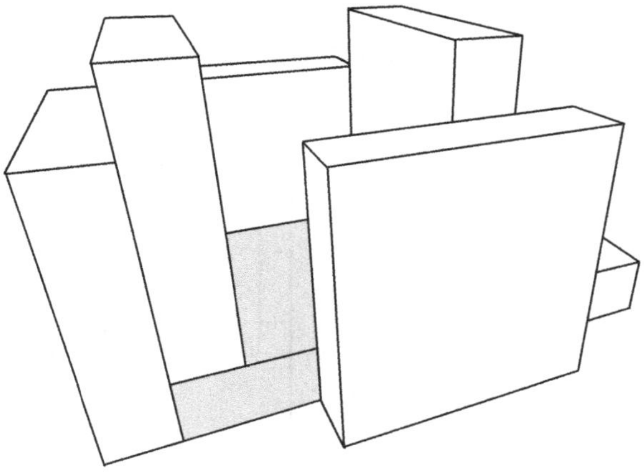

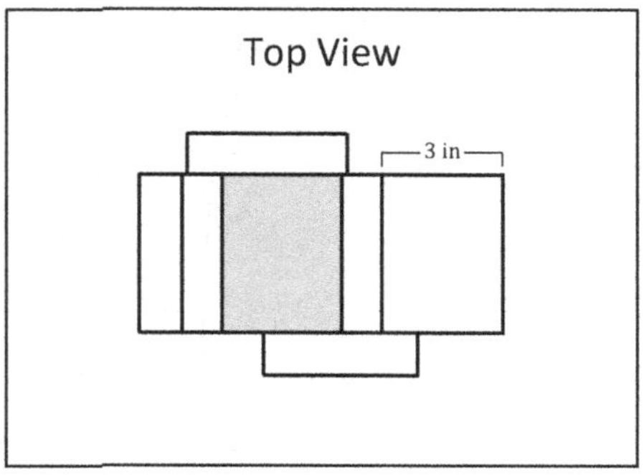

Count: ____________________

12.

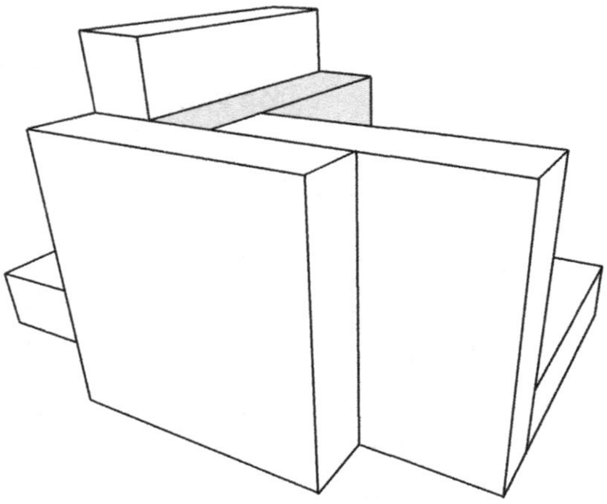

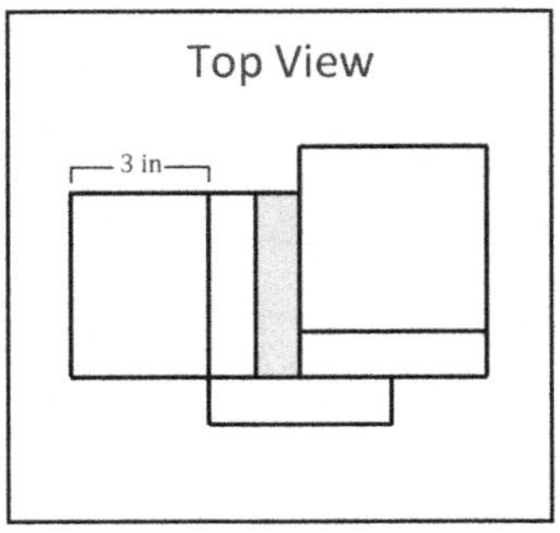

Count: ______________________

13.

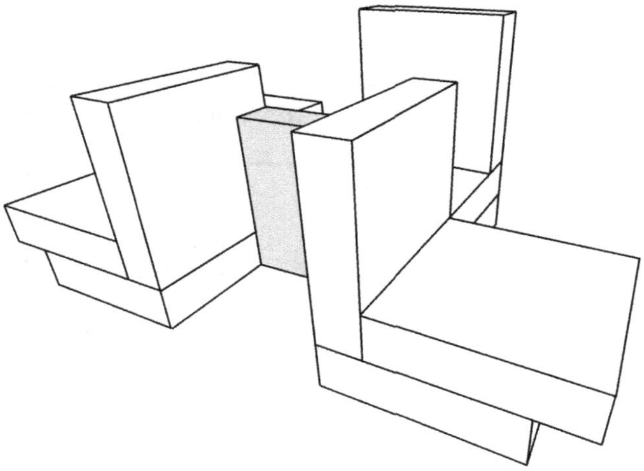

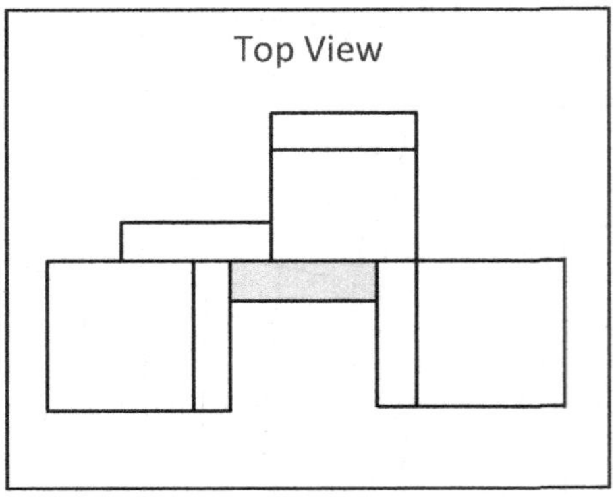

Count: ______________________

14.

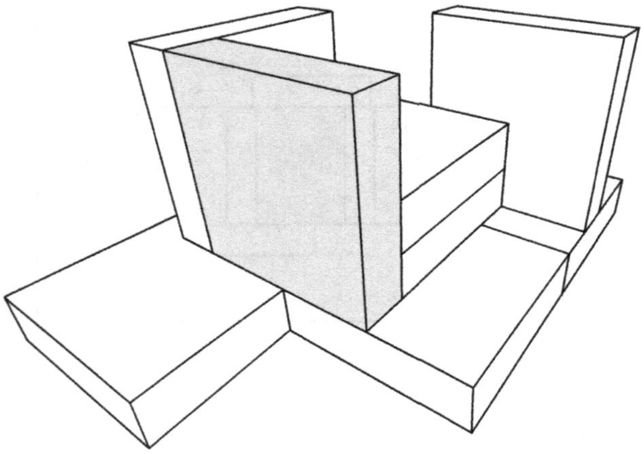

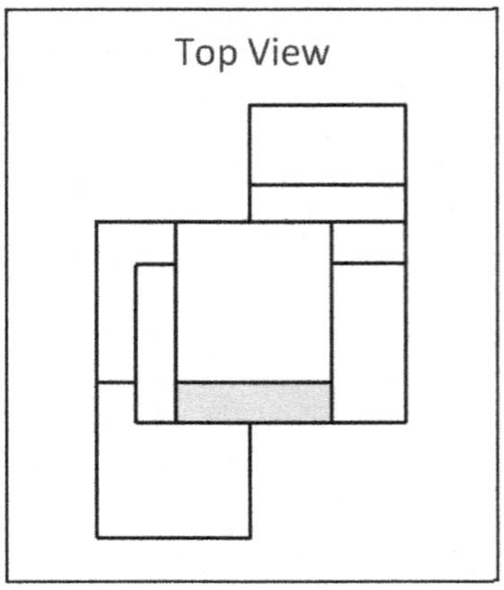

Count: ______________________

15.

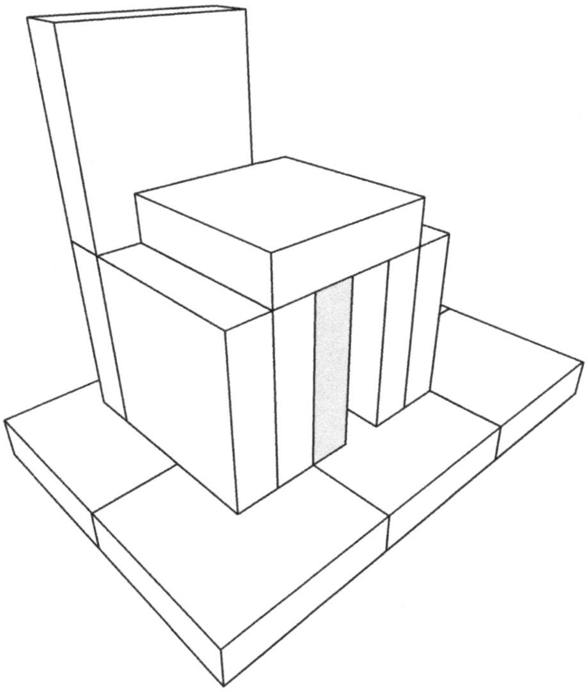

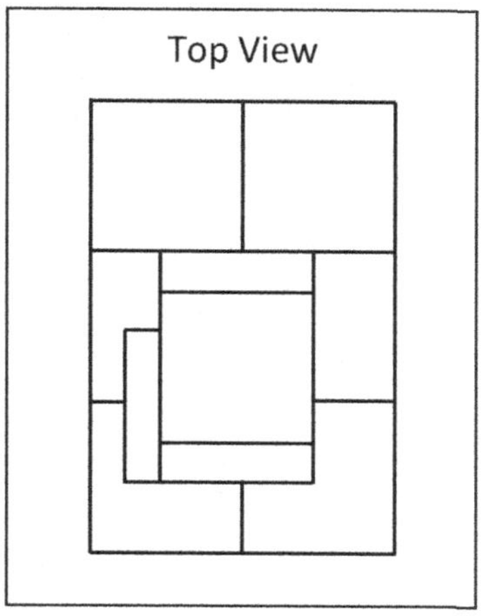

Count: ______________________

16.

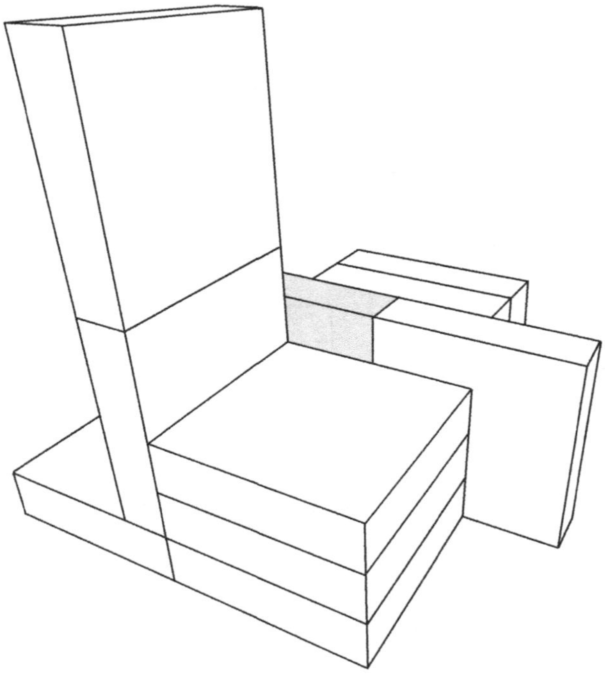

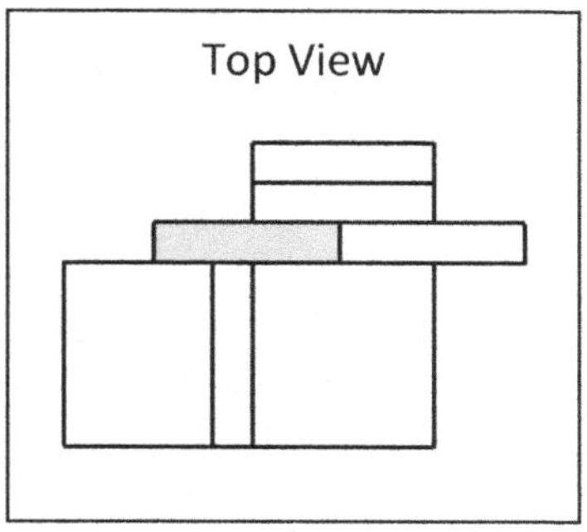

Count: ______________________

17. 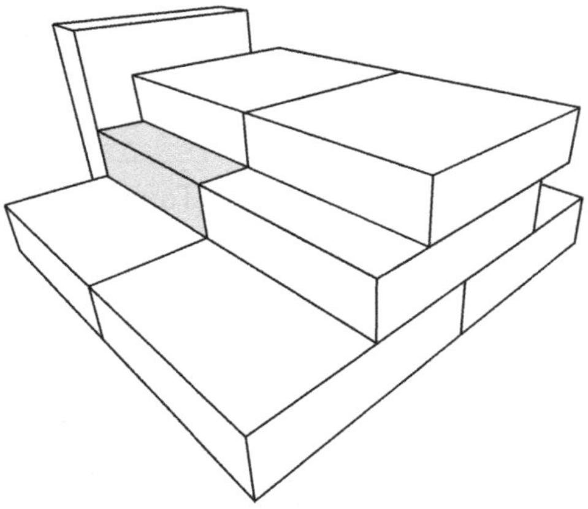

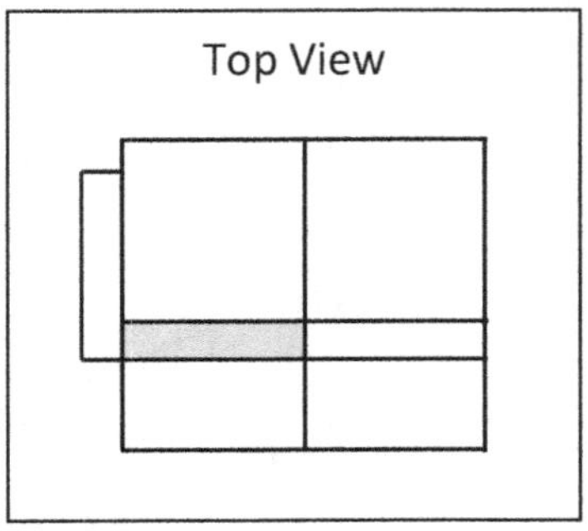

Count: ____________________

18. 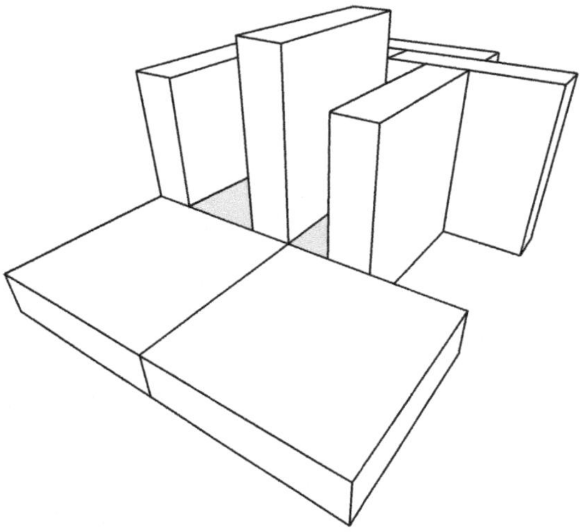

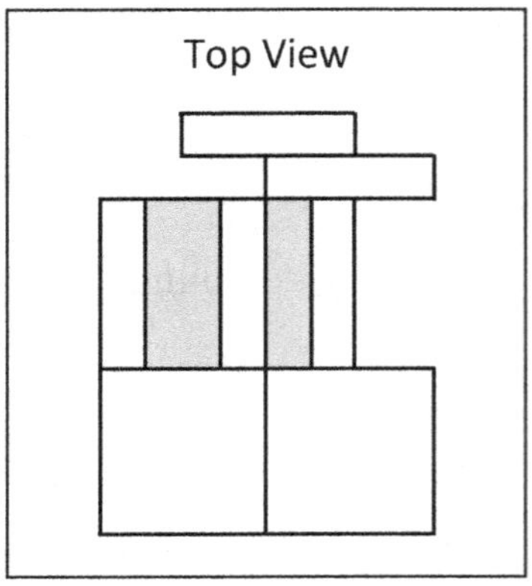

Count: ____________________

19. 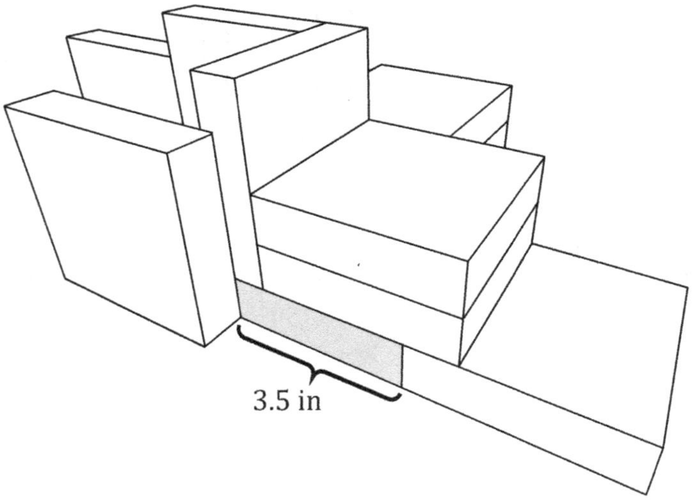

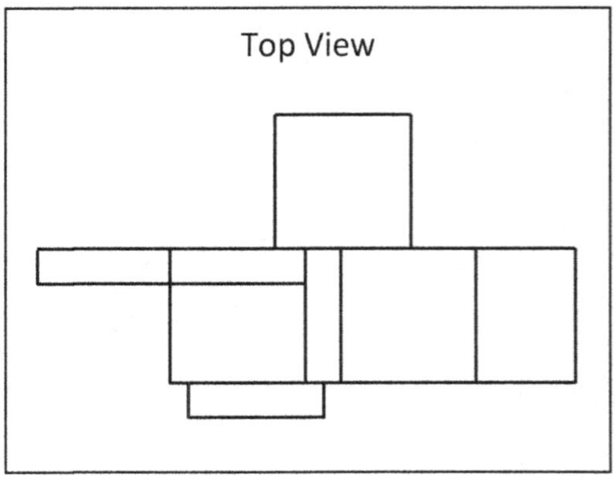

Count: ____________________

20.

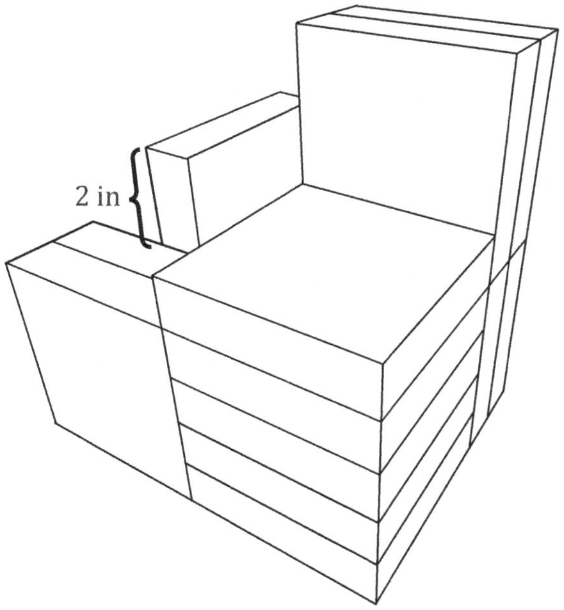

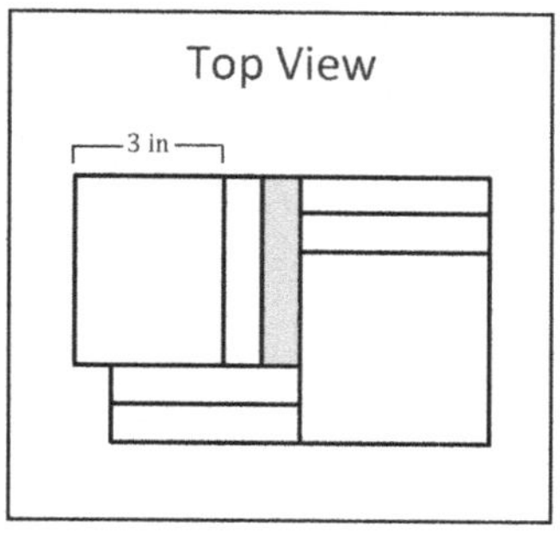

Count: ____________________

21.

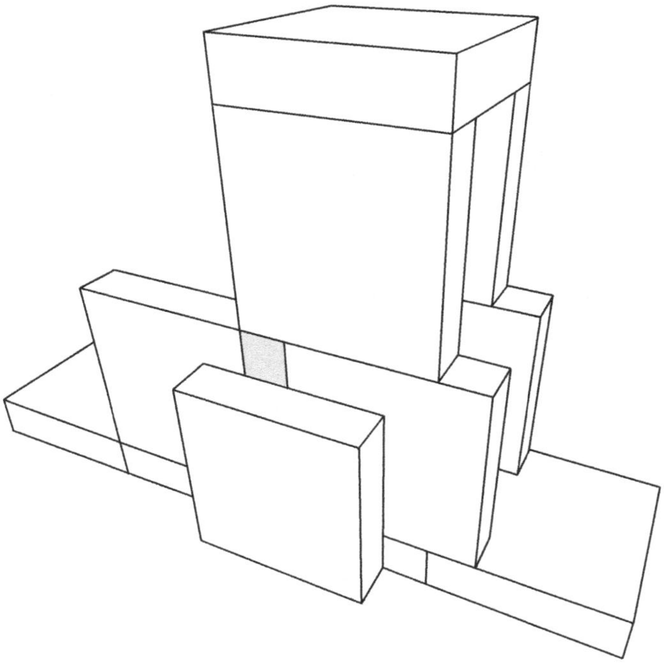

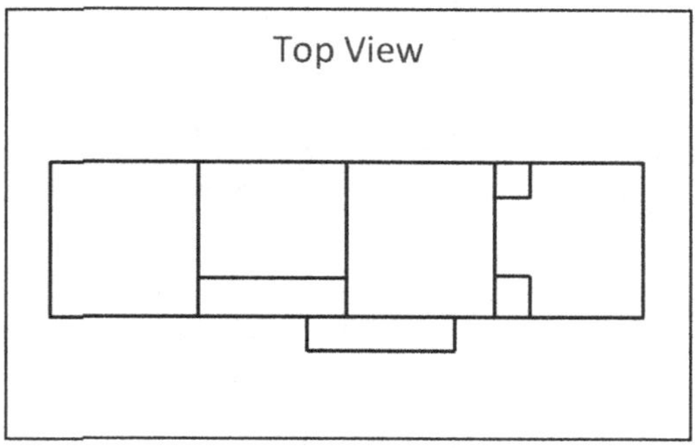

Count: ____________________

22.

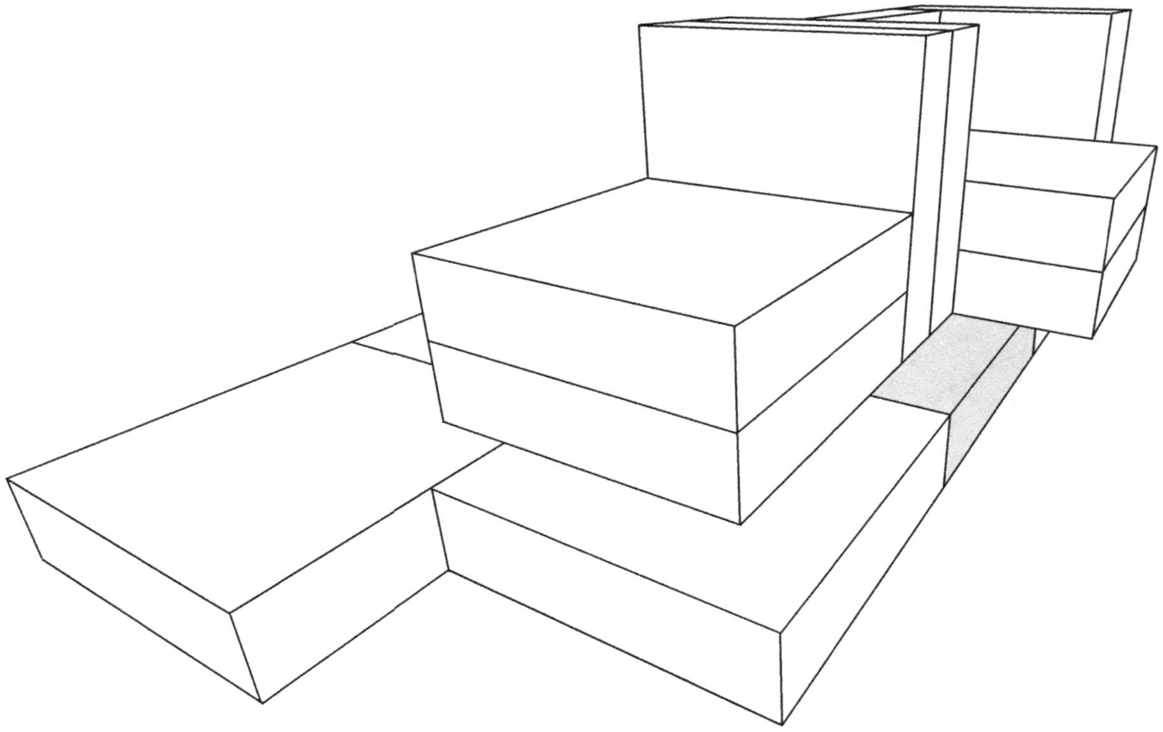

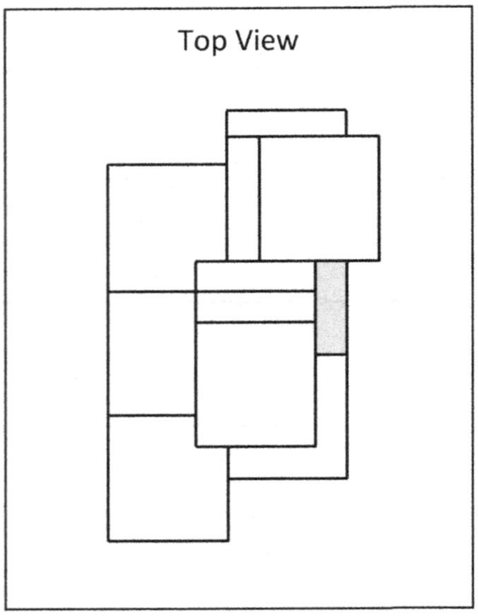

Count: ______________________

Counting Cubes

Count the number of cubes in each pile. Every cube in the pile is the same size and shape. Some of the cubes are hidden from view. No cube is floating completely in the air; if a cube is not resting on the ground, there must be another cube underneath holding it in place. For some questions, the top view will be provided, showing a "bird's eye view" of the pile of cubes.

1.

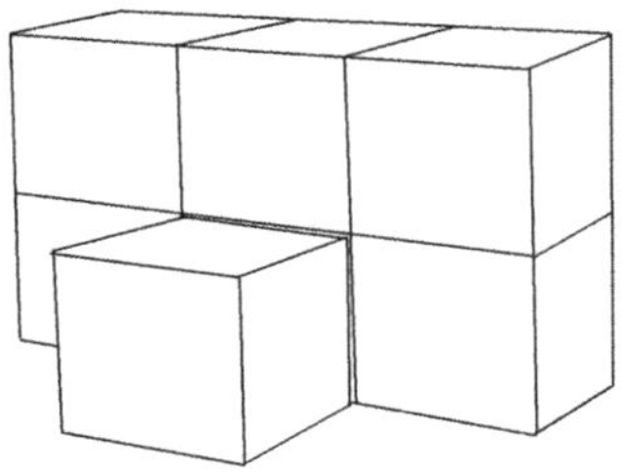

Count: ______________________

2.

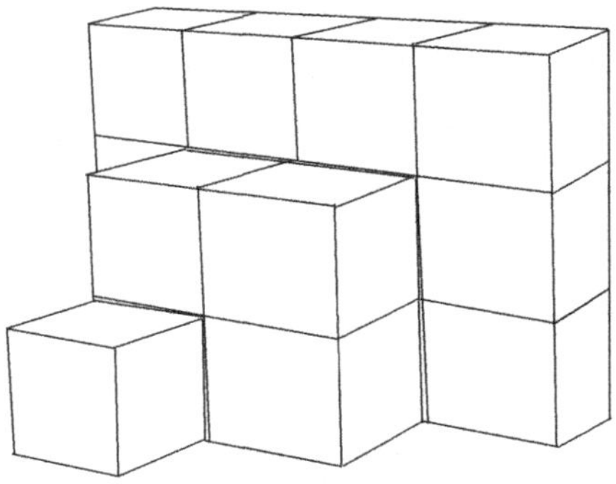

Count: ______________________

3.

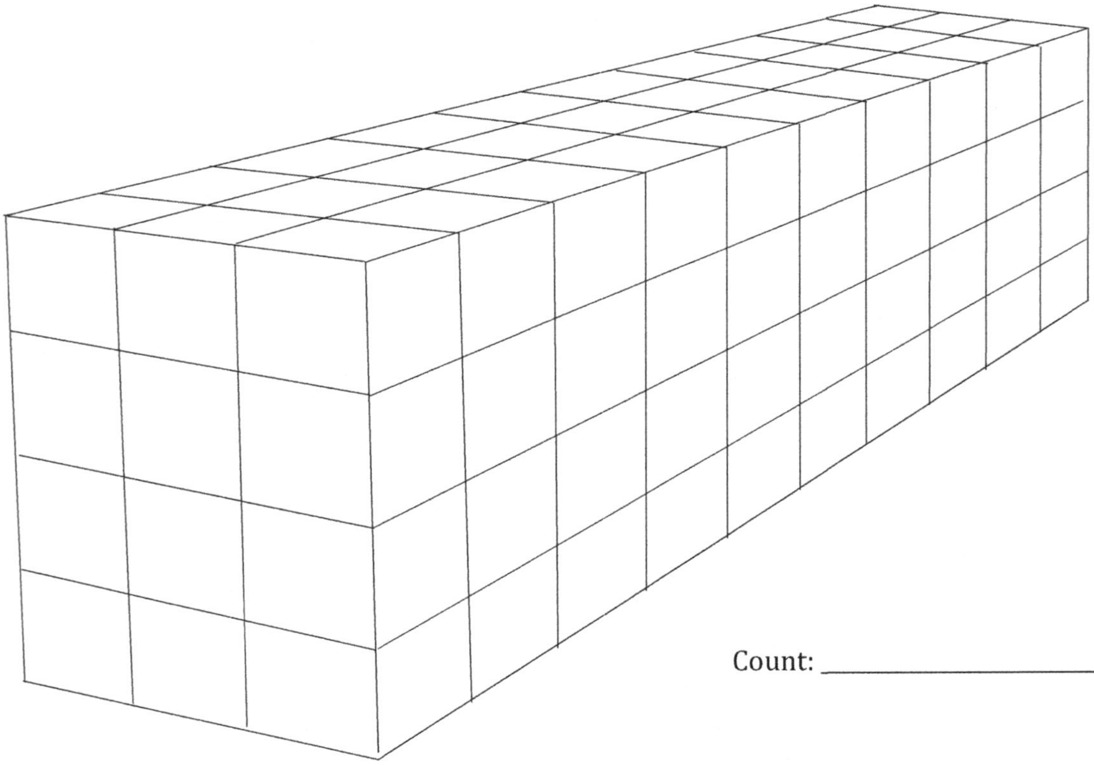

Count: ______________________

4.

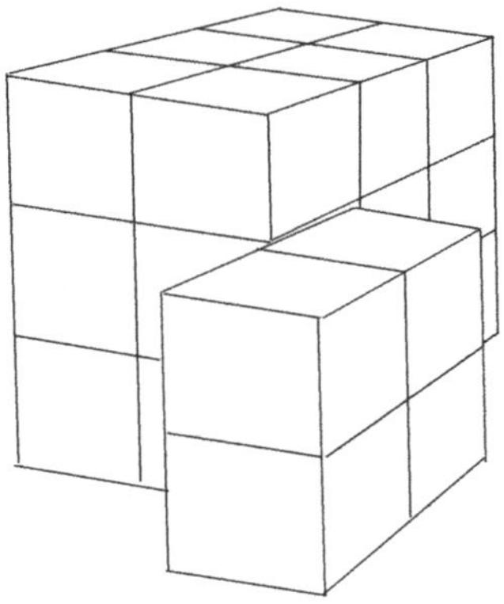

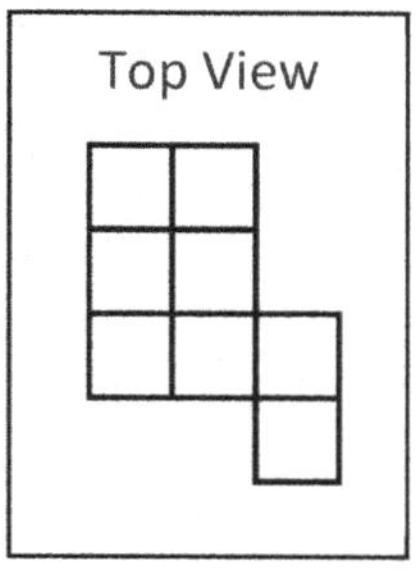

Count: ____________________

5.

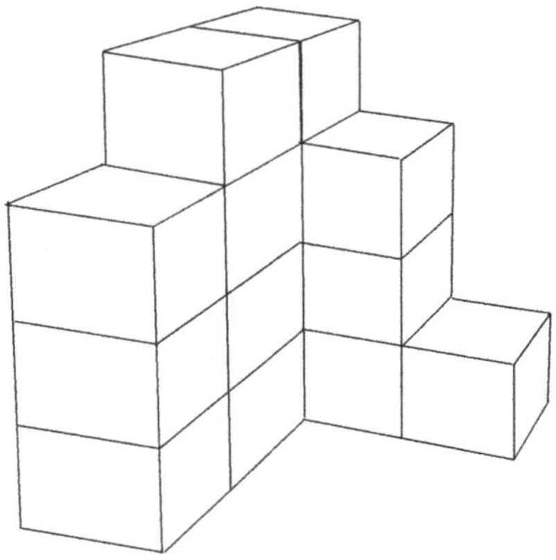

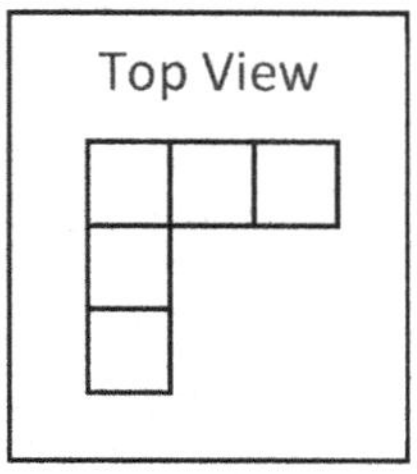

Count: ____________________

6.

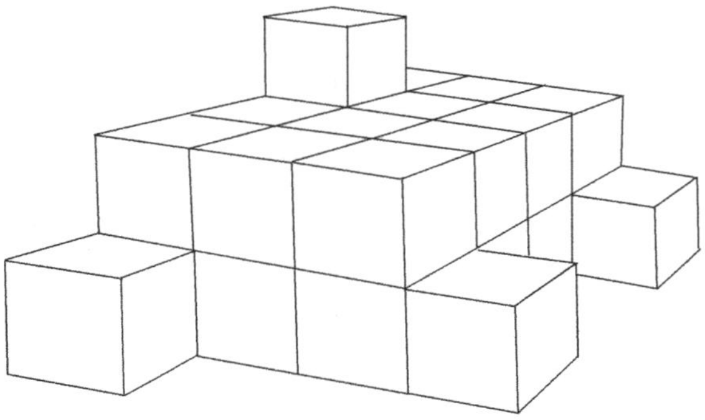

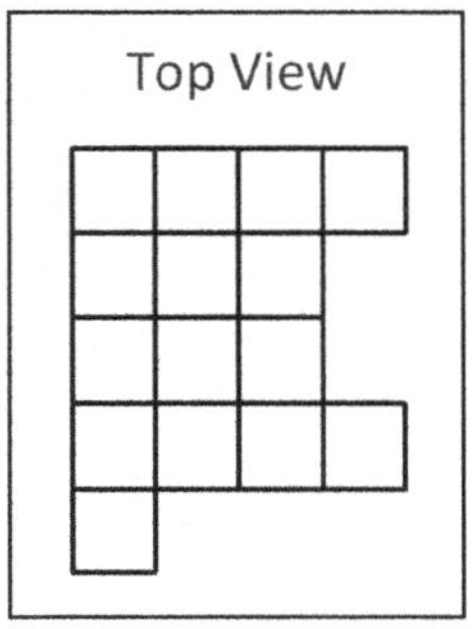

Count: ____________________

7.

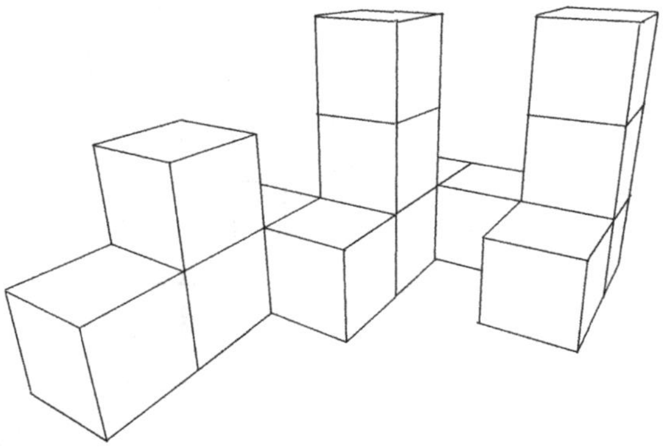

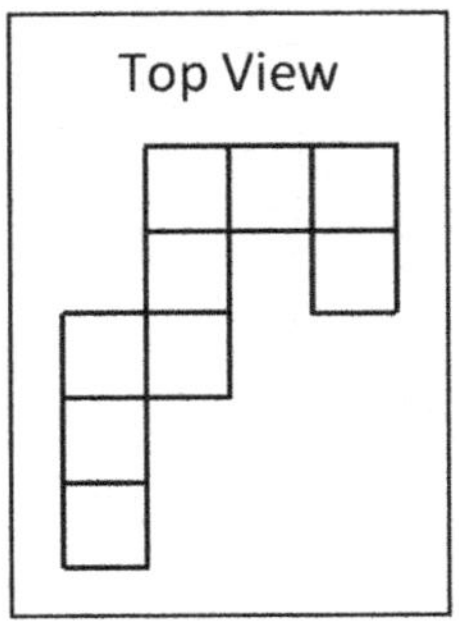

Count: ______________________

8.

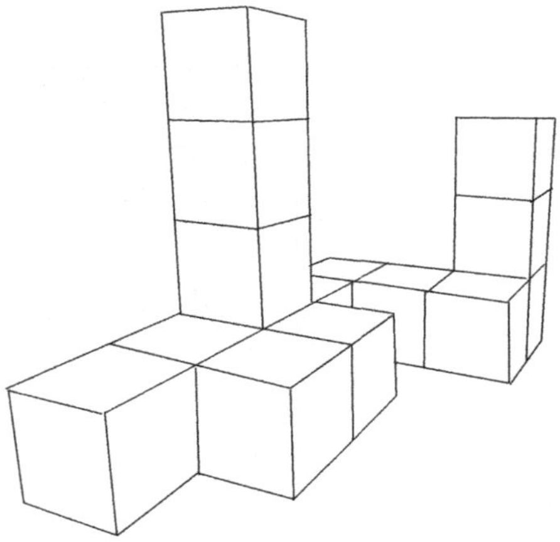

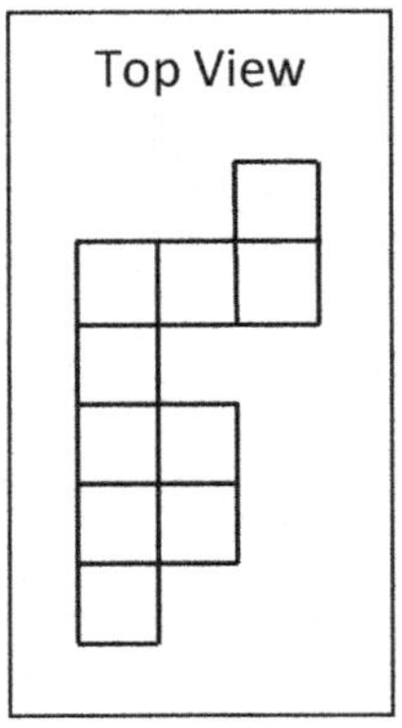

Count: ______________________

9.

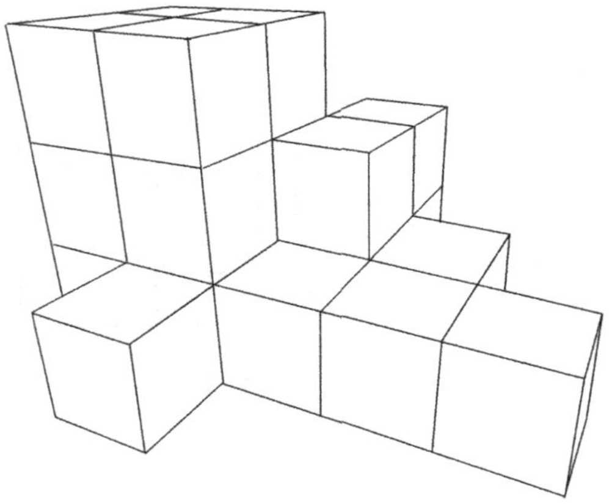

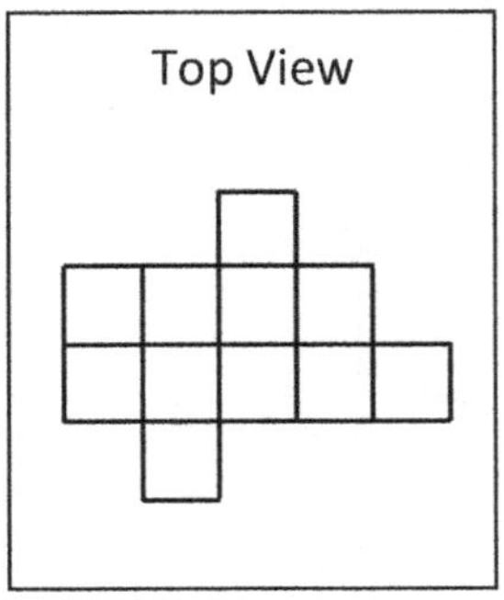

Count: ______________________

10.

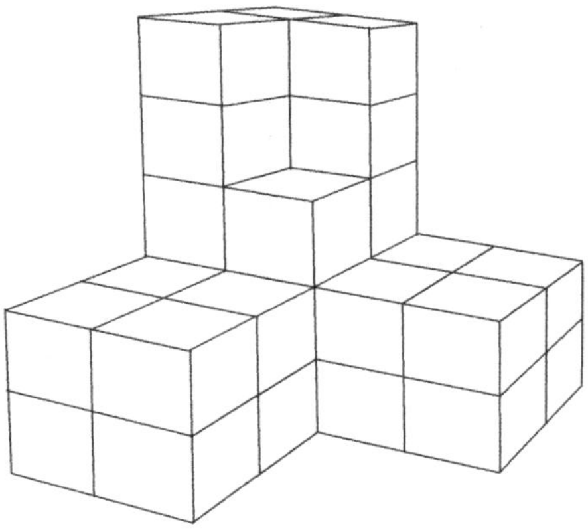

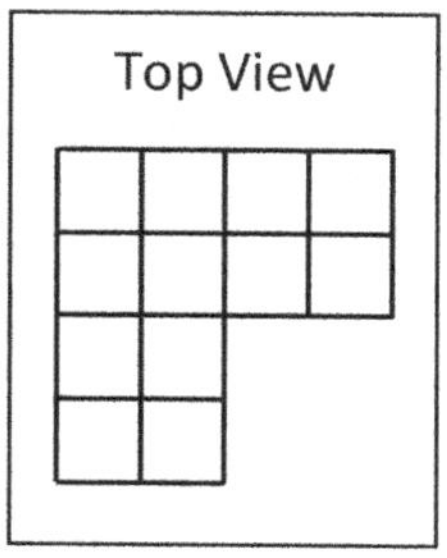

Count: ______________________

11.

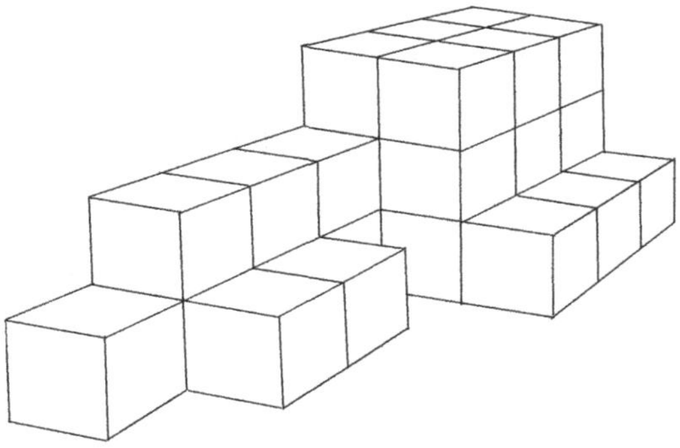

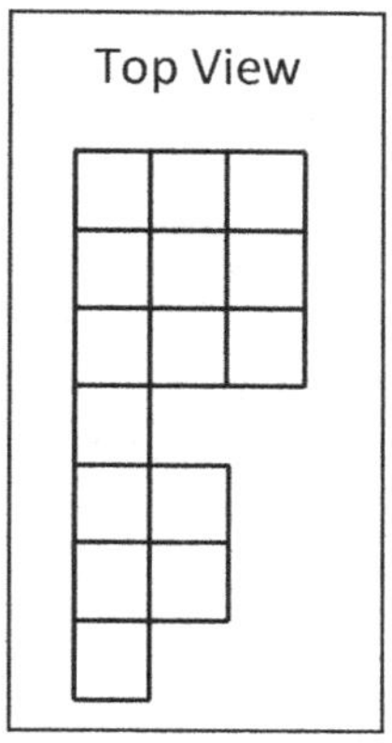

Count: ______________________

12.

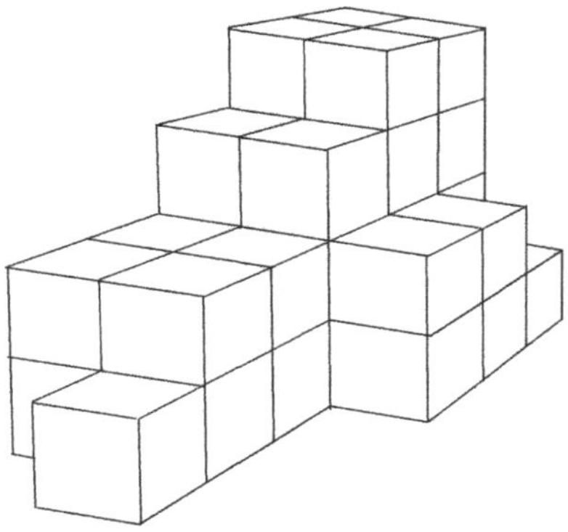

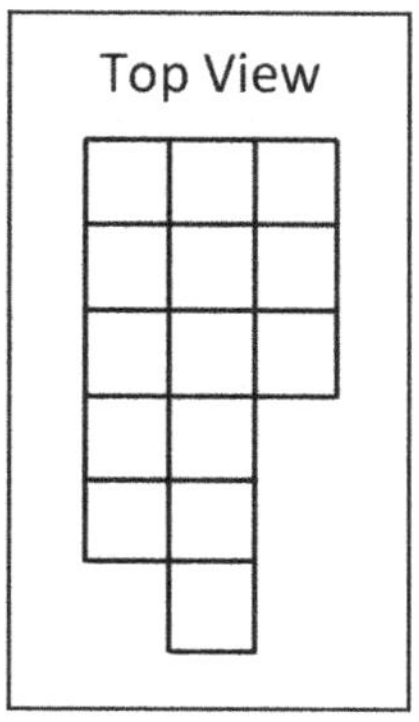

Count: ______________________

13.

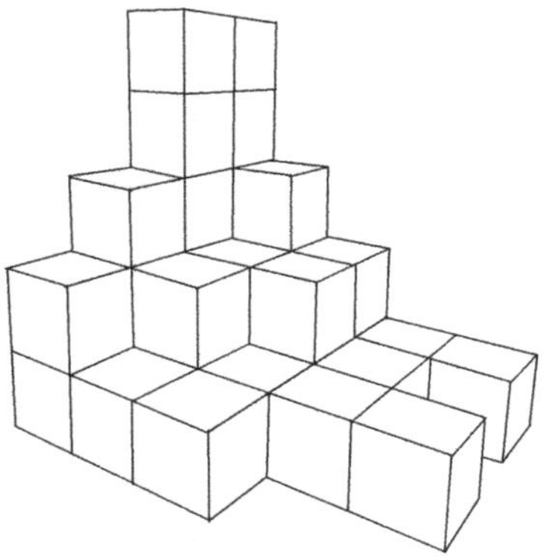

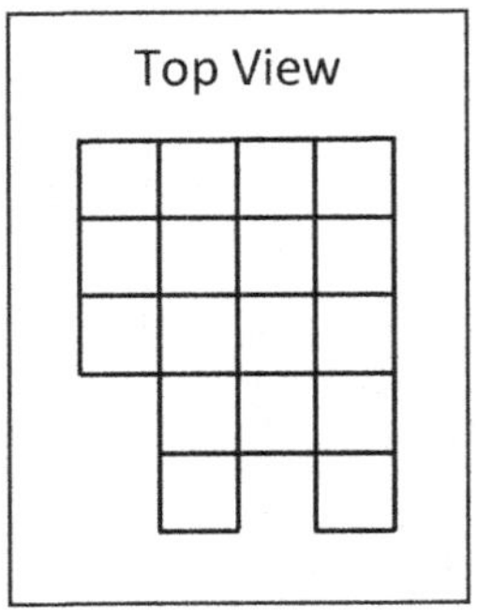

Count: ____________________

14.

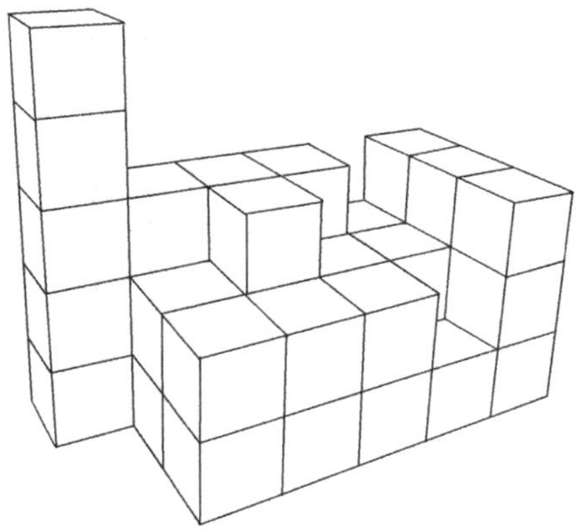

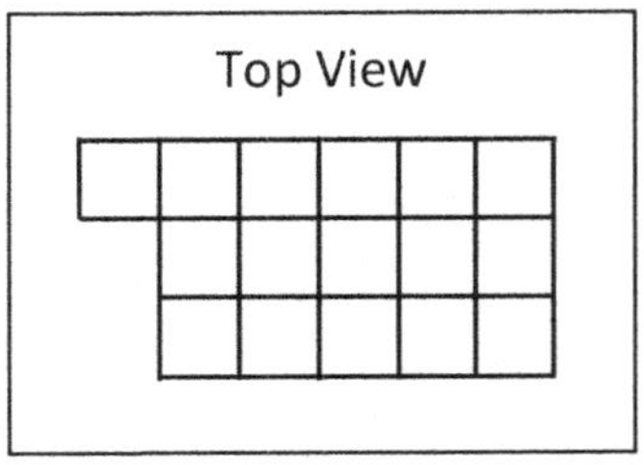

Count: ____________________

15.

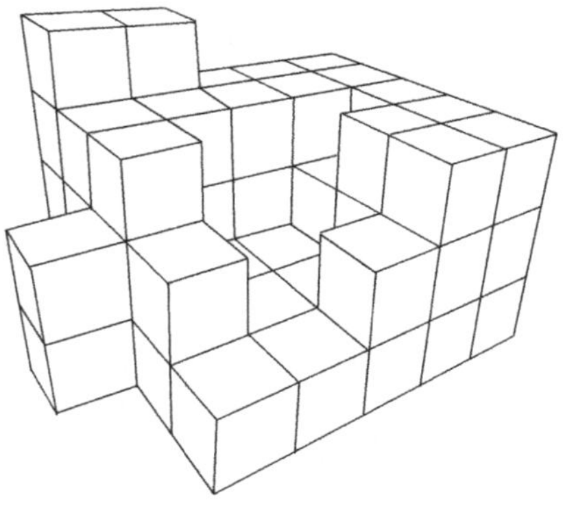

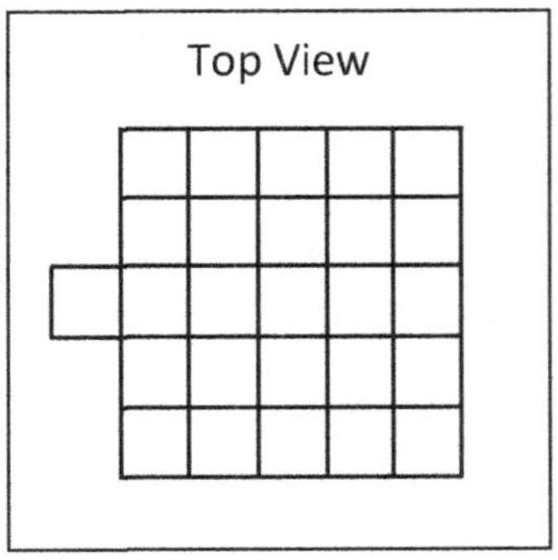

Count: ____________________

16.

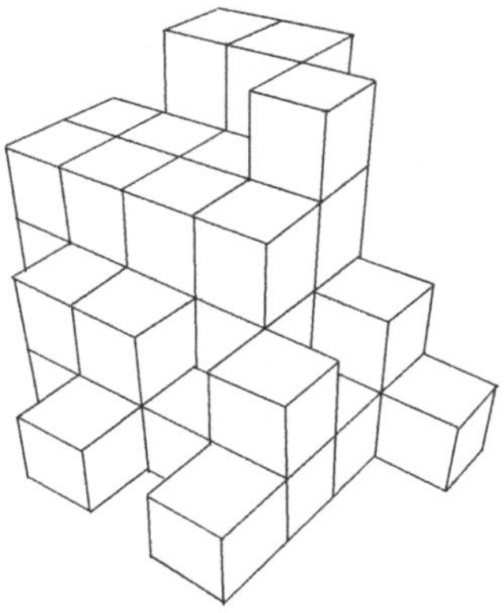

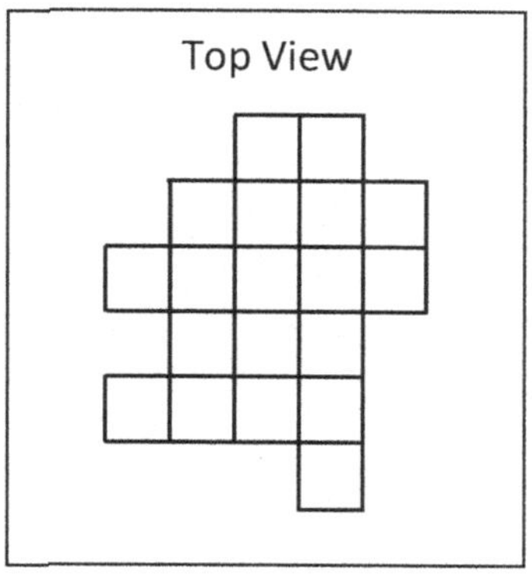

Count: ______________________

17.

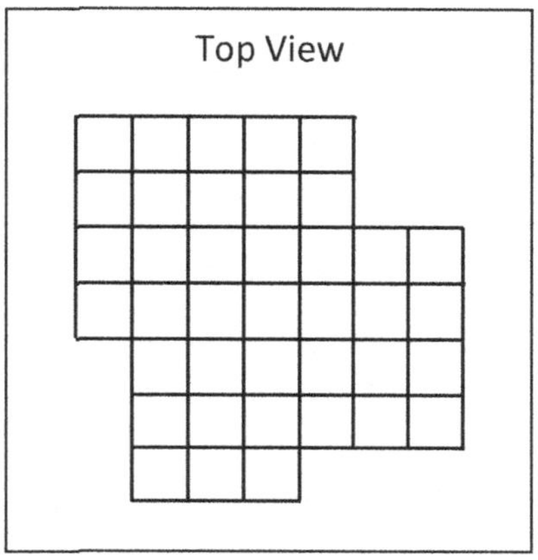

Count: ______________________

18.

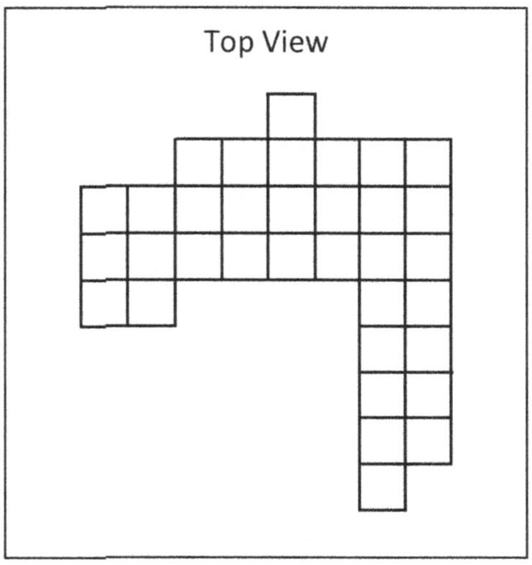

Count: ______________________

19.

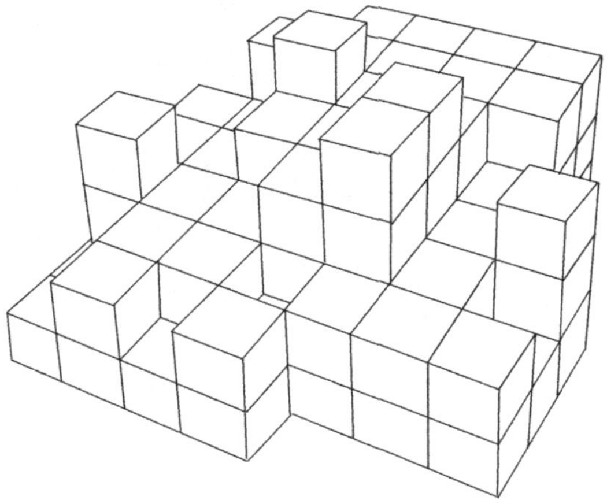

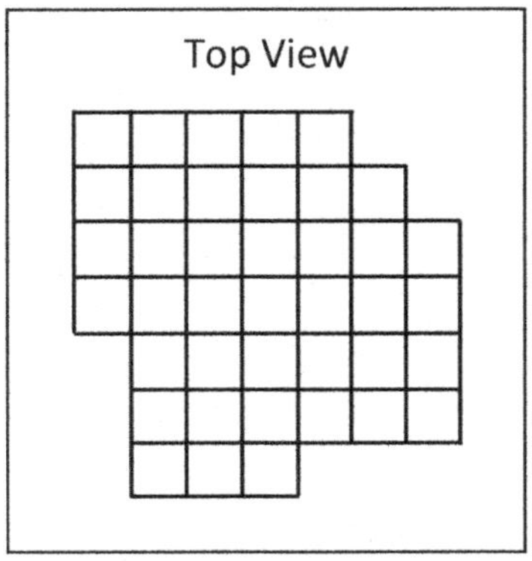

Count: ____________________

20.

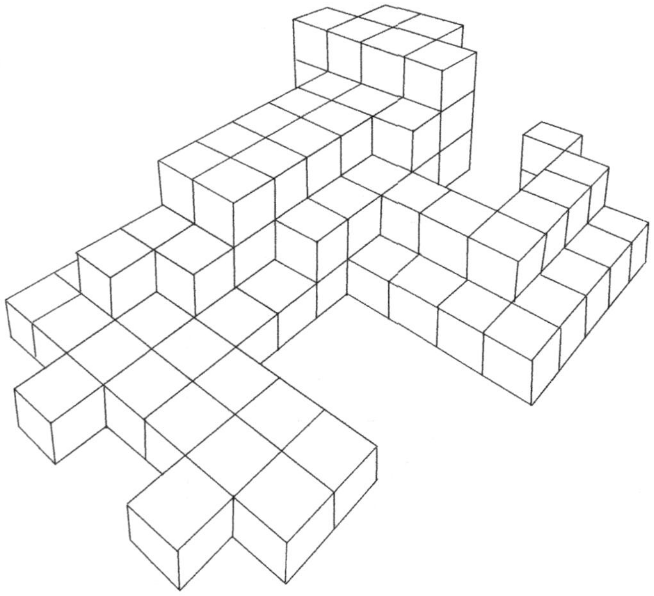

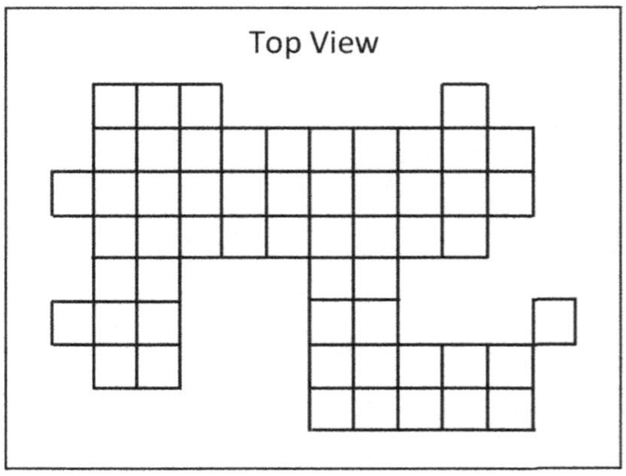

Count: ____________________

21.

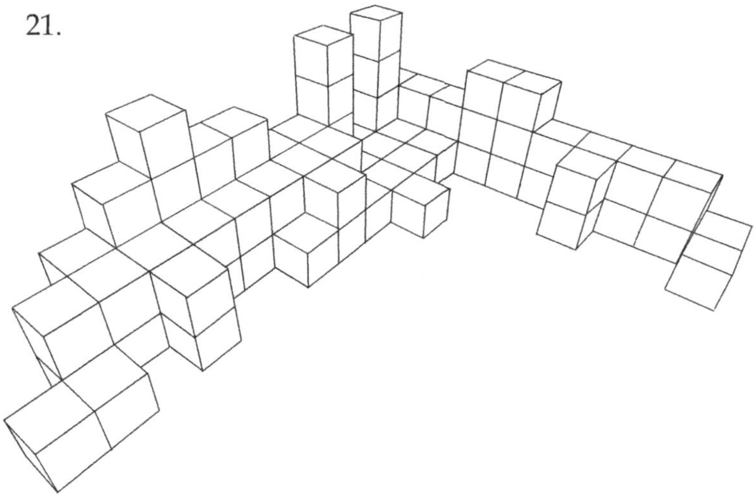

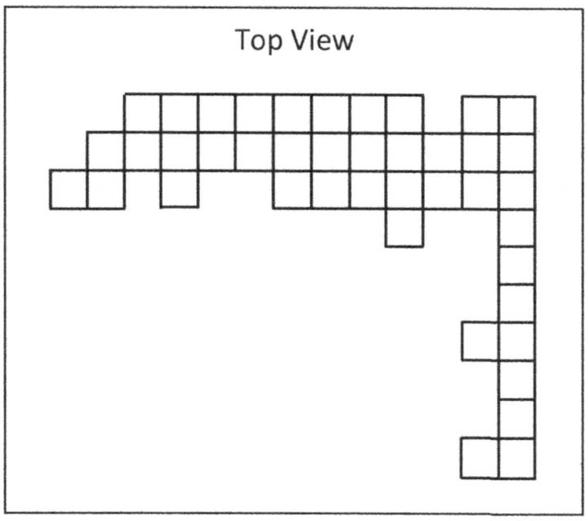

Count: ______________________

22.

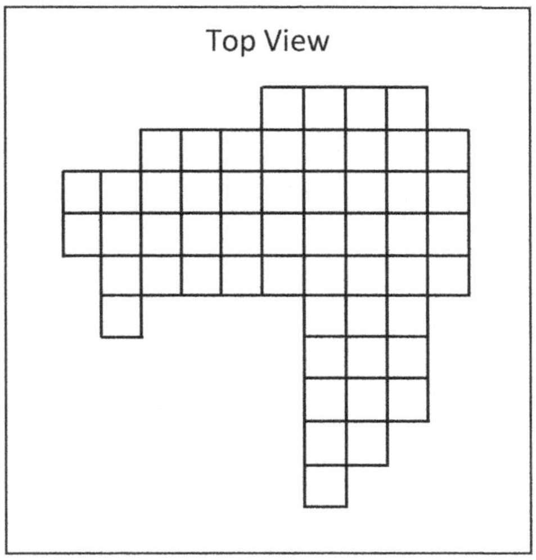

Count: ______________________

23.

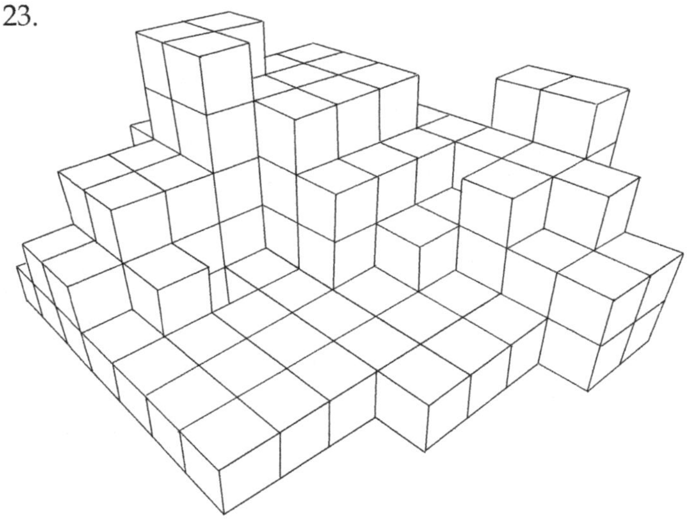

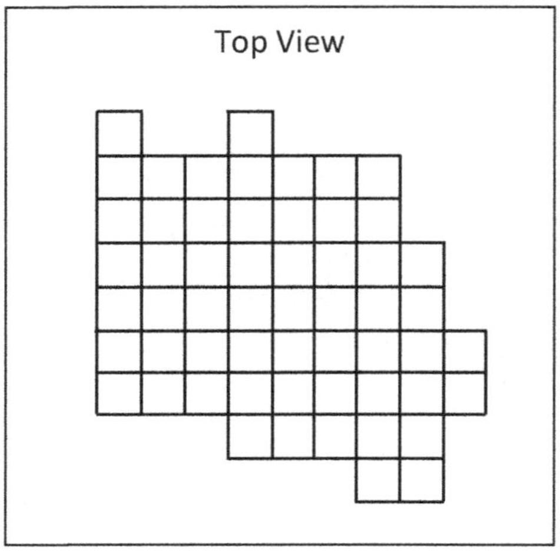

Count: ______________________

24.

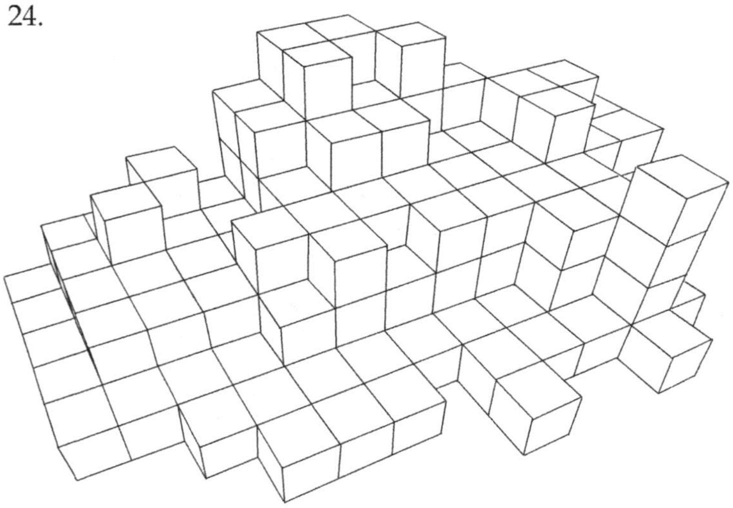

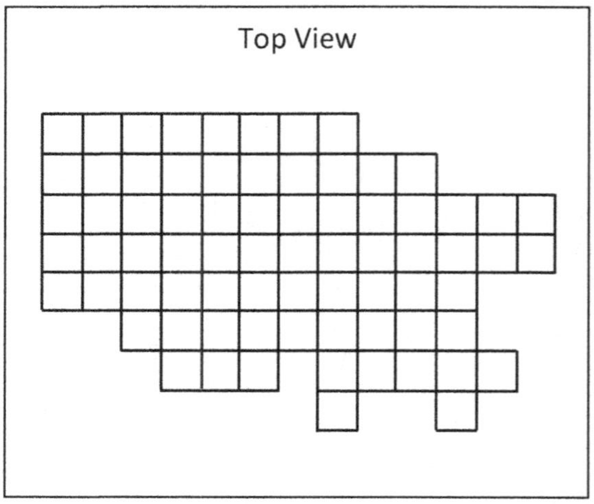

Count: ______________________

Connecting Cubes

State the **least** number of cubes needed to connect the two shaded cubes.

Example:

Answer: 5

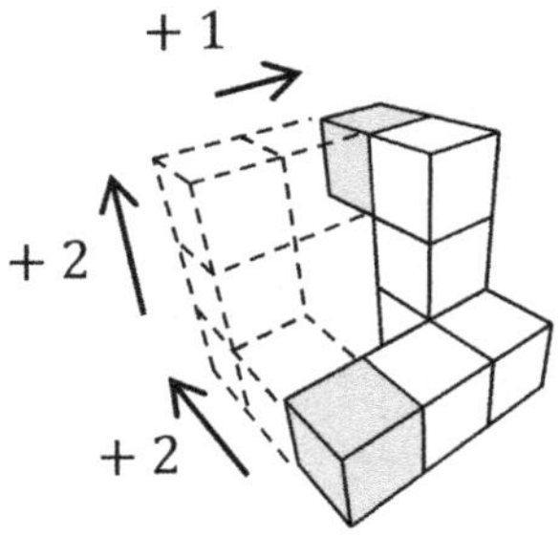

1.

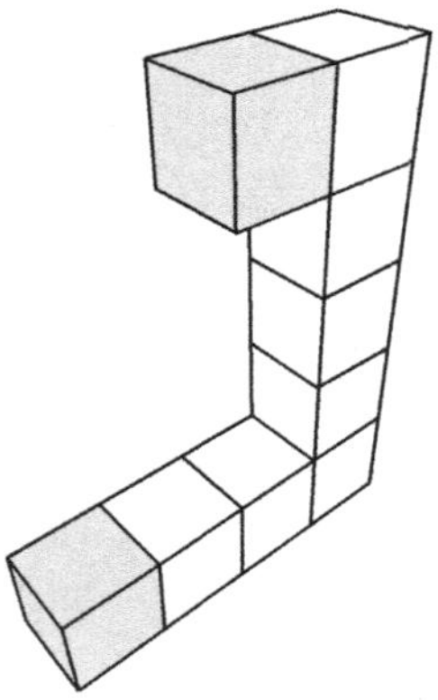

____

2.

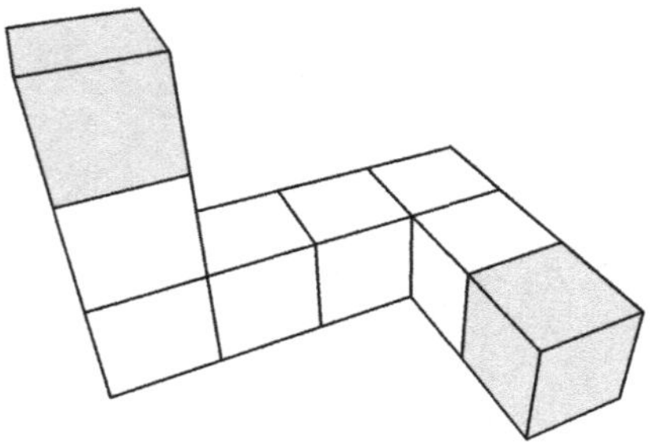

____

3.

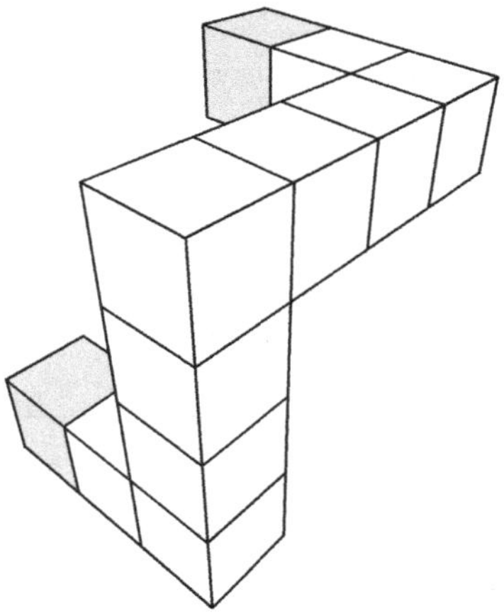

____

4.

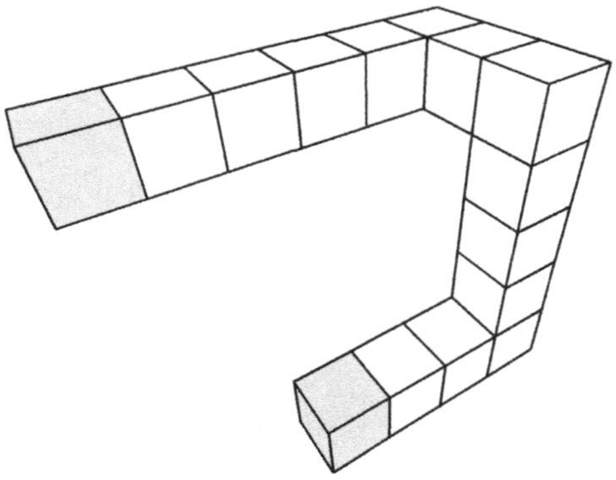

____

5.

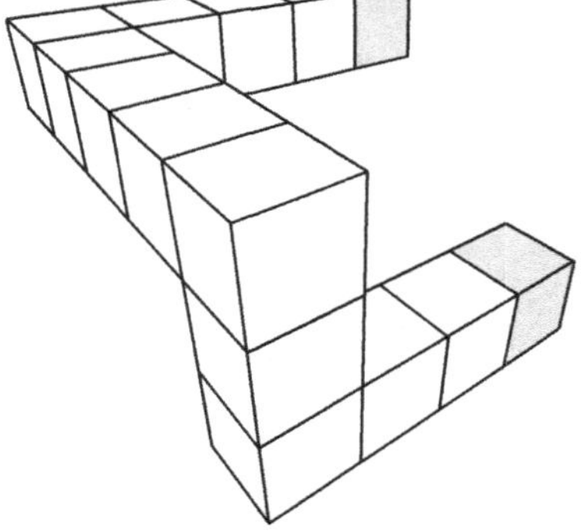

____

6.

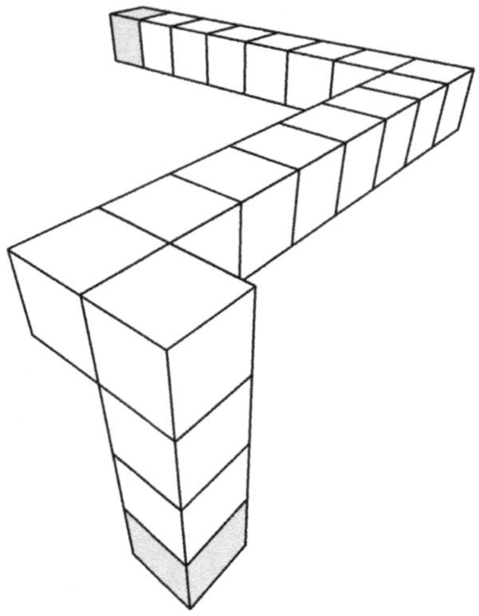

____

7.

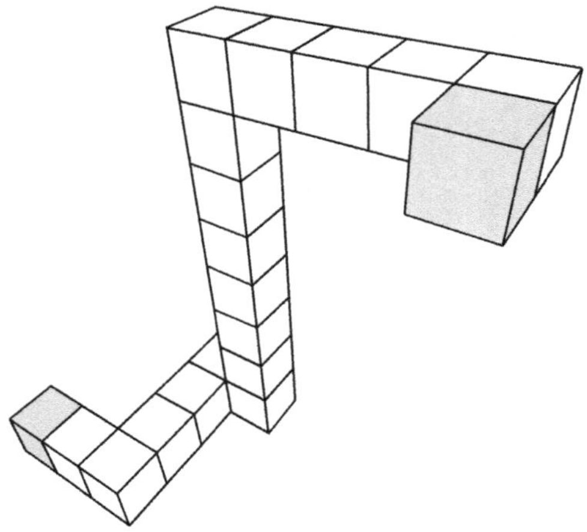

____

8.

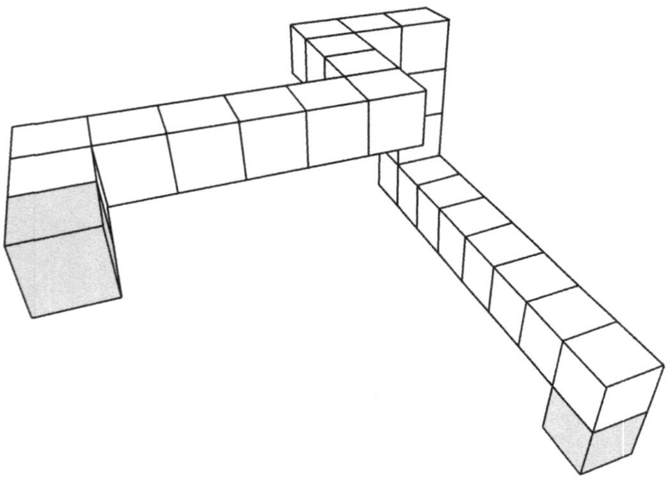

____

9.

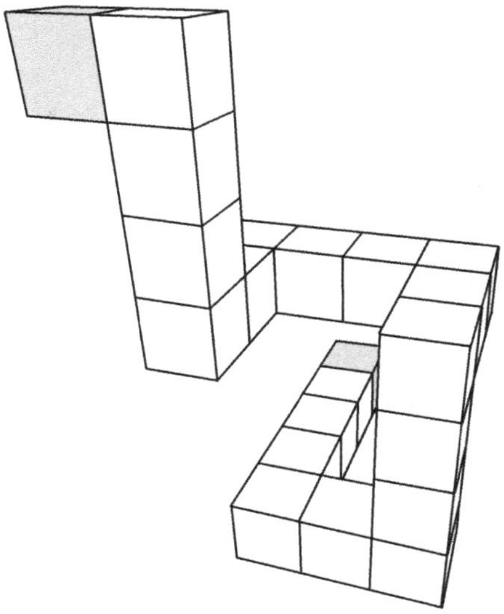

10.

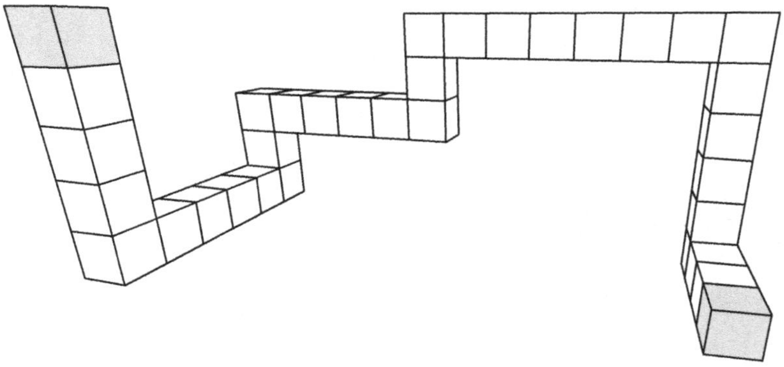

Top View, Front View, and Side View

Part 1

Choose the image that matches the specified view.

Note: Some of the cubes are hidden from view. No cube is floating in the air, if a cube is not resting on the ground, there must be another cube underneath holding it in place.

1.

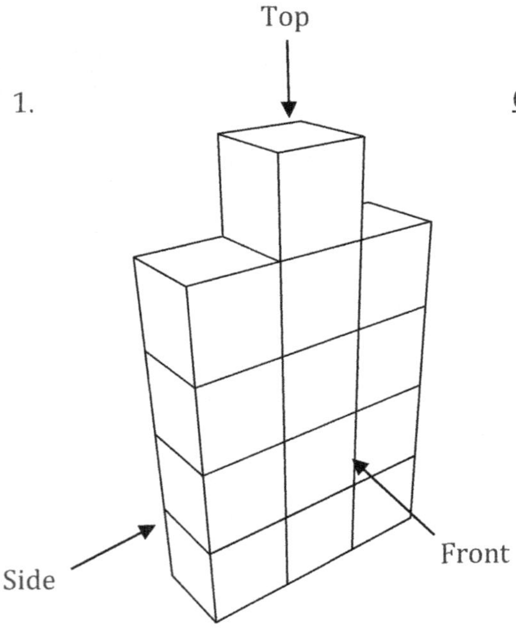

<u>Circle the Side View</u>

a. 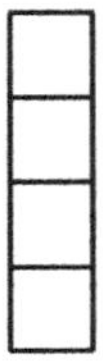

b. 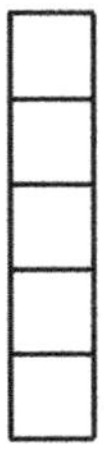

c. 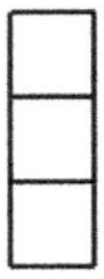

d. 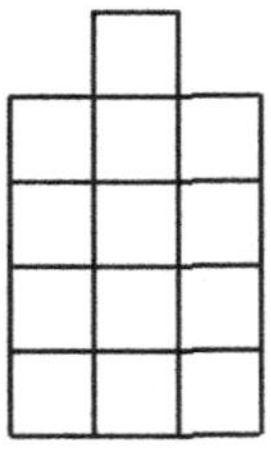

2.

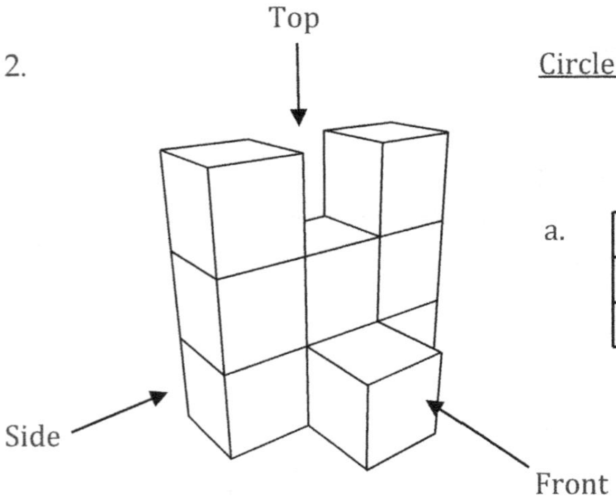

<u>Circle the Front View</u>

a. 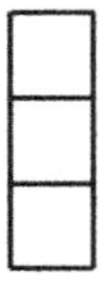

b. 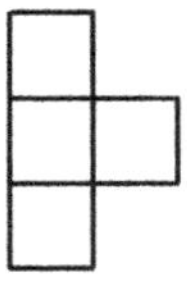

c. 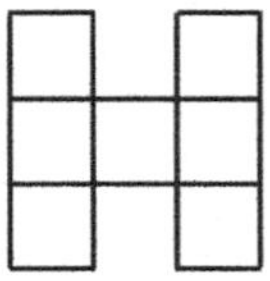

d. 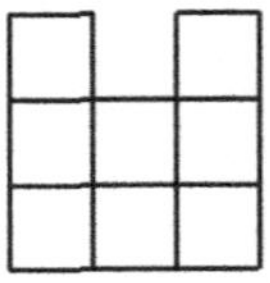

3.

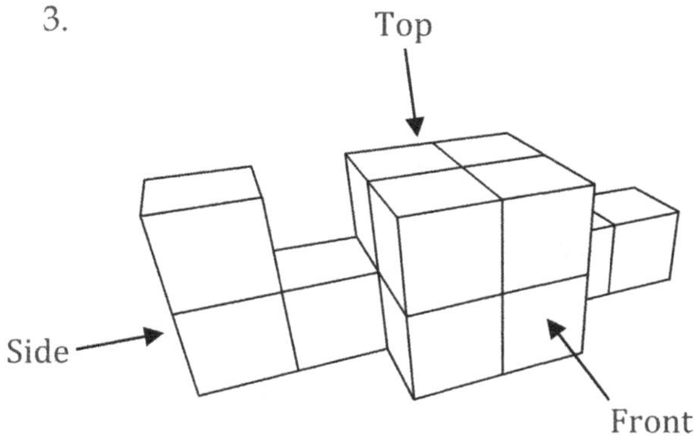

Circle the Top View

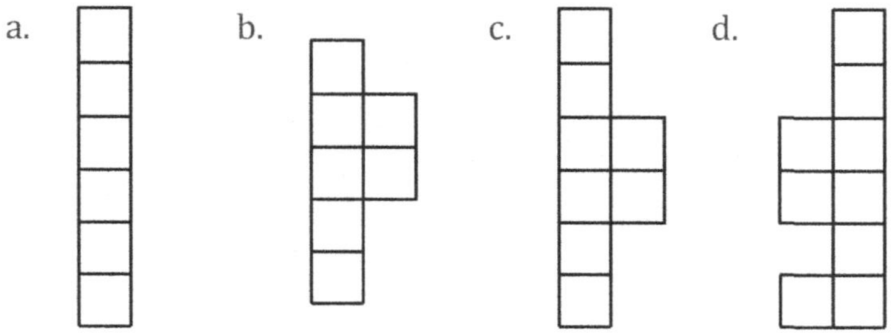

4.

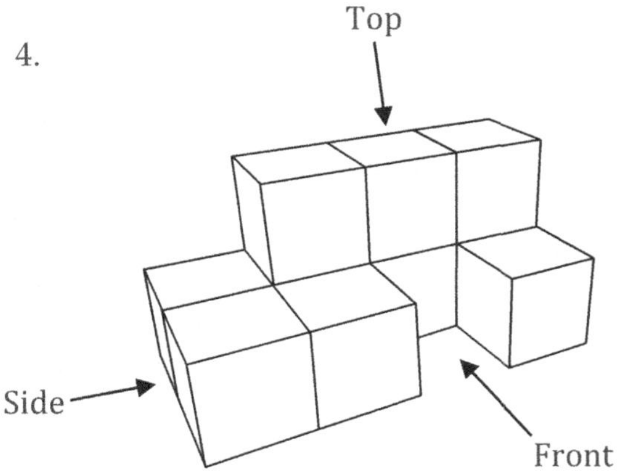

Circle the Side View

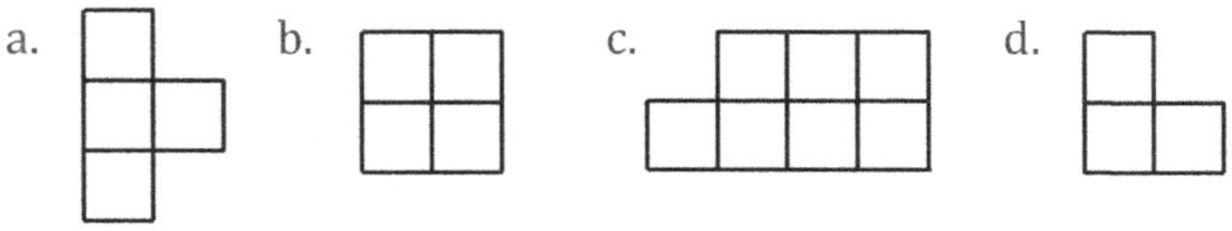

5.

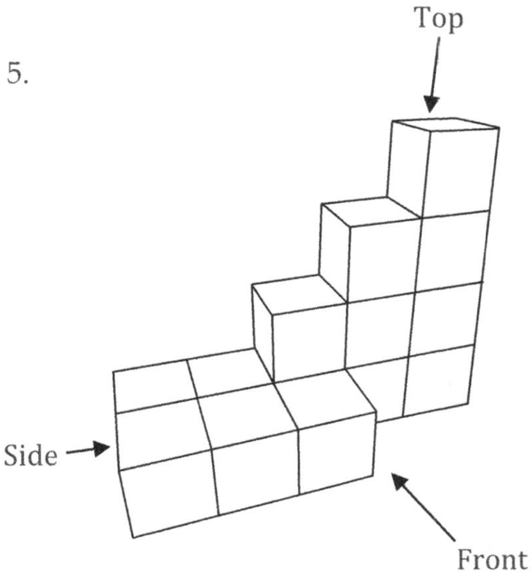

Circle the Front View

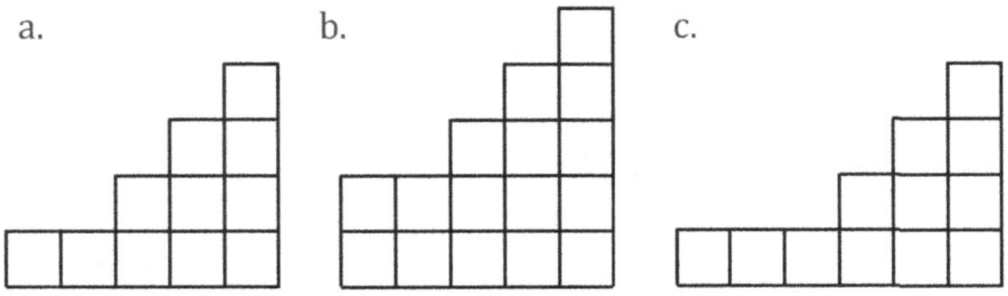

6.

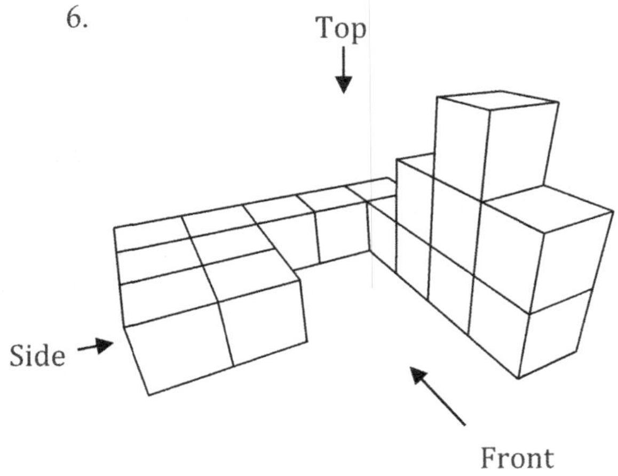

Circle the Top View

a. 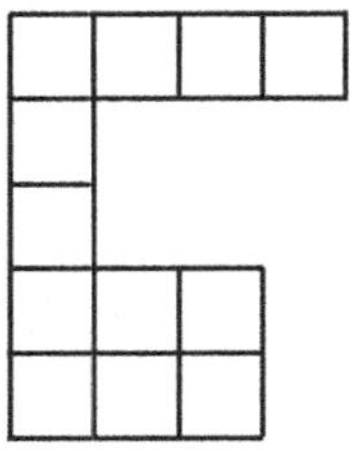

b. 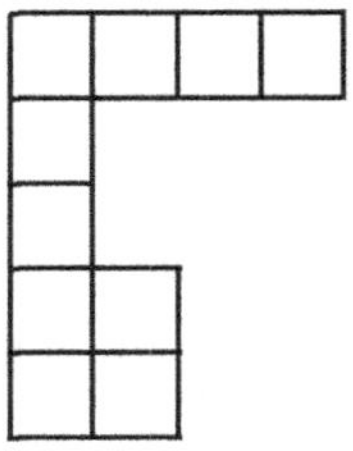

c. 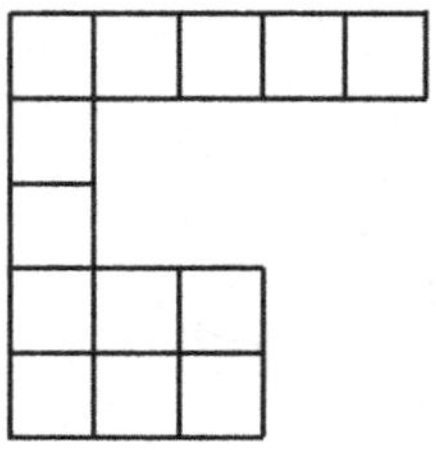

7.

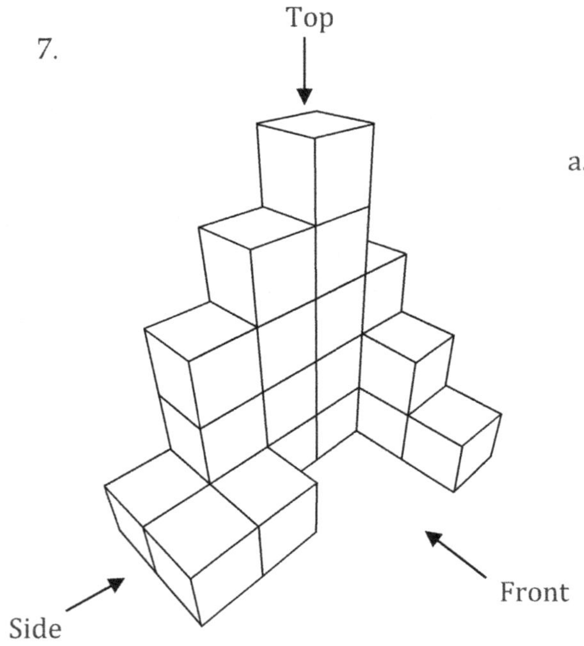

Circle the Top View

a. 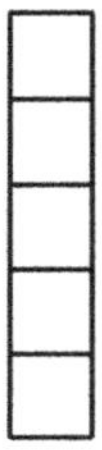

b. 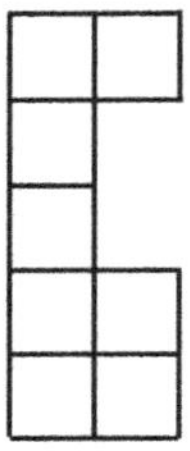

c. 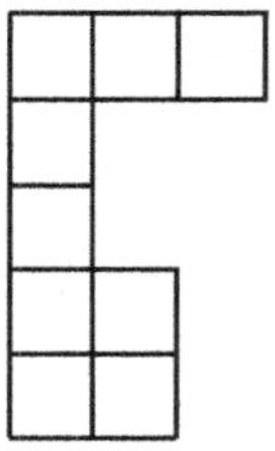

d. 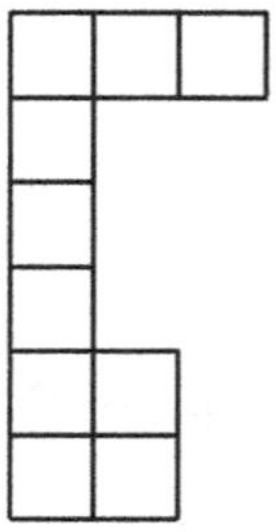

8.

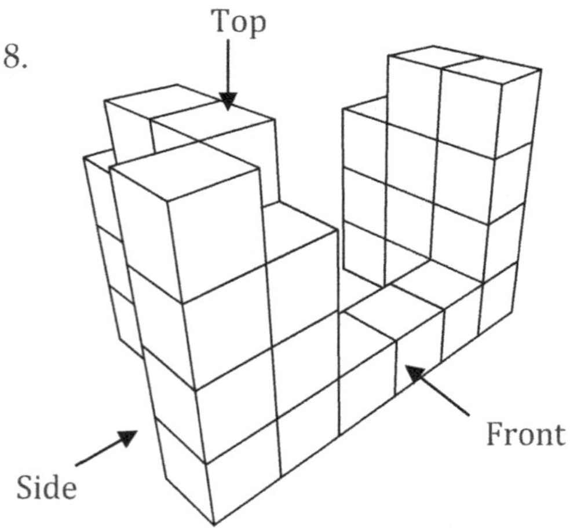

Circle the Side View

a. 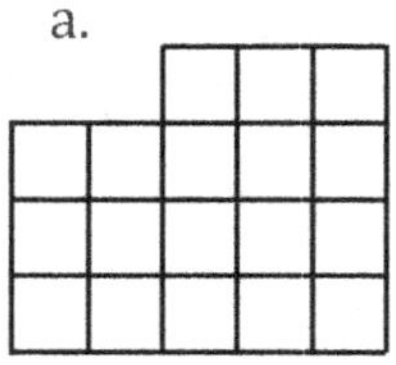

b. 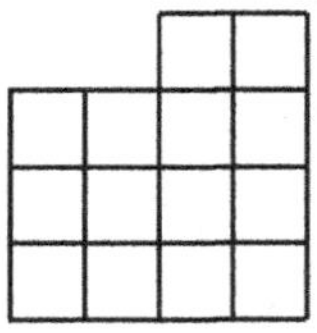

c. 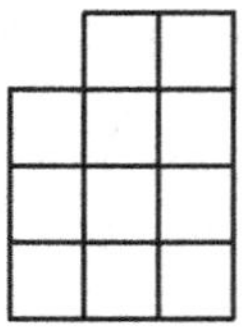

d. 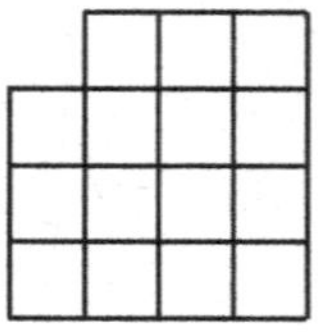

9.

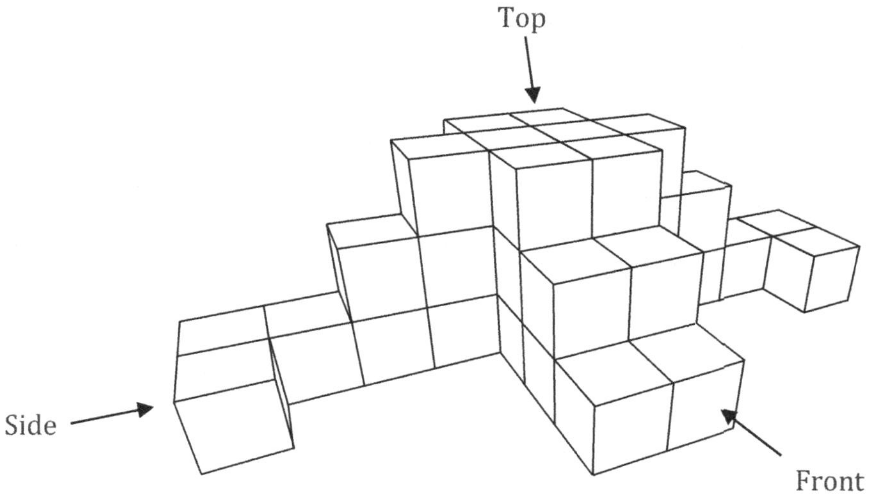

Circle the Front View

a.

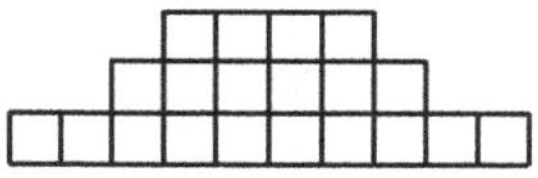

b.

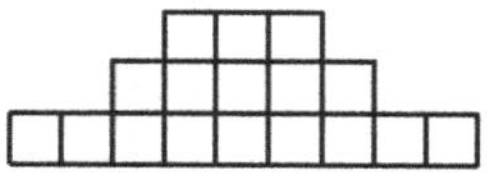

c.

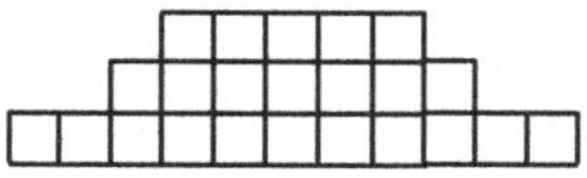

d.

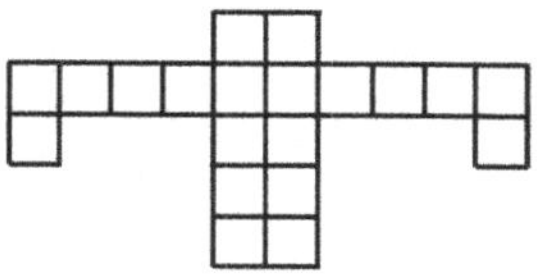

10.

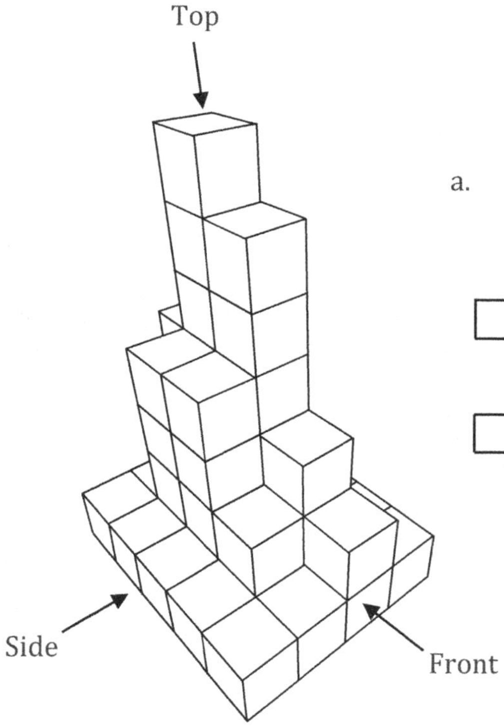

Circle the Side View

a.

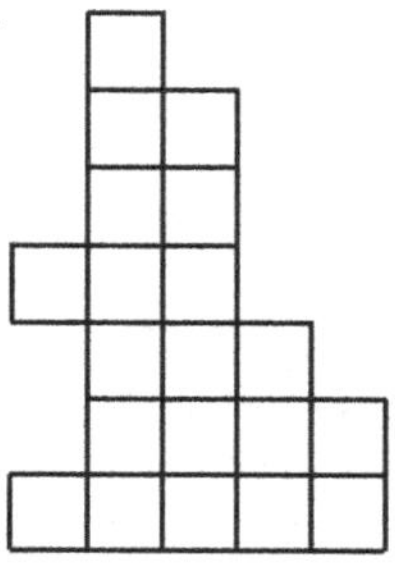

b.

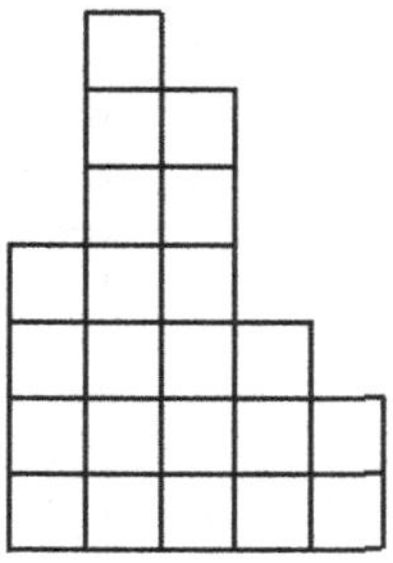

c.

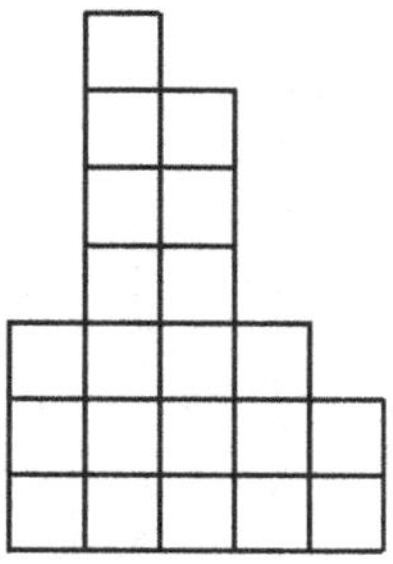

d.

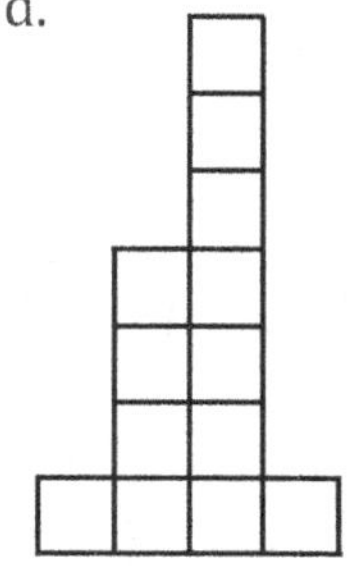

Top View, Front View, and Side View

Part 2

Shade in the squares that would indicate the correct front view, top view, and side view.

Note: Some of the cubes are hidden from view. No cube is floating in the air, if a cube is not resting on the ground, there must be another cube underneath holding it in place. The exact number of cubes found in each pile will be provided.

Example

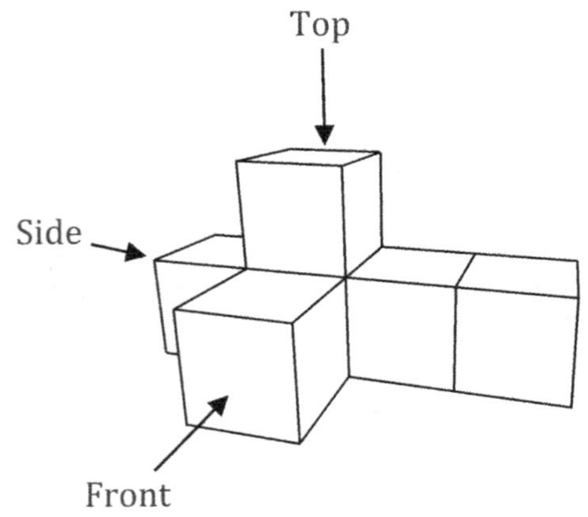

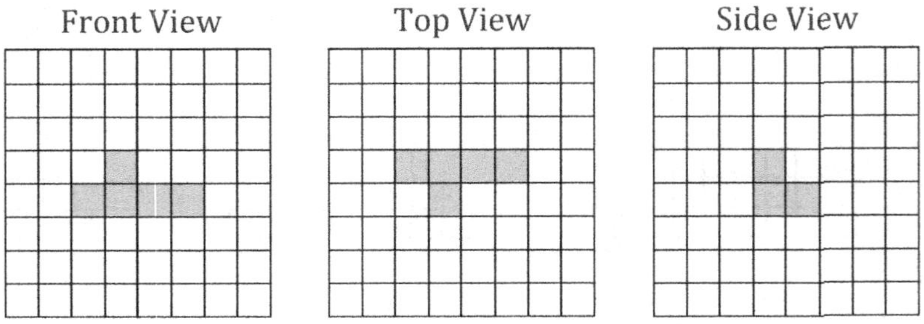

1.

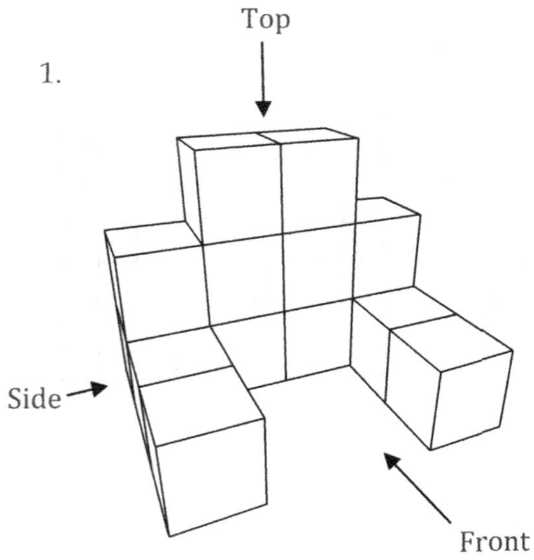

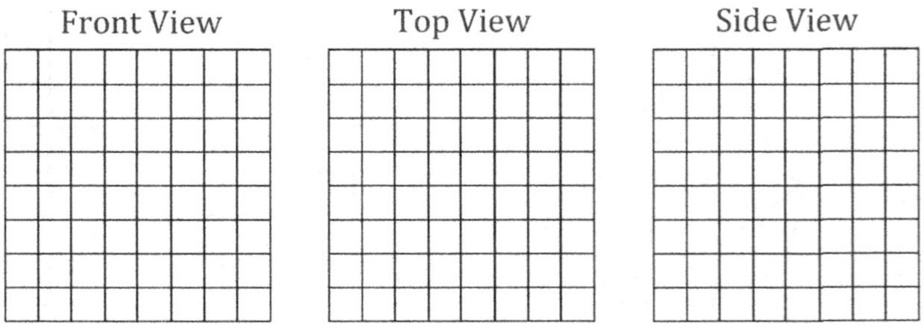

Total number of cubes in the pile: 14

2.

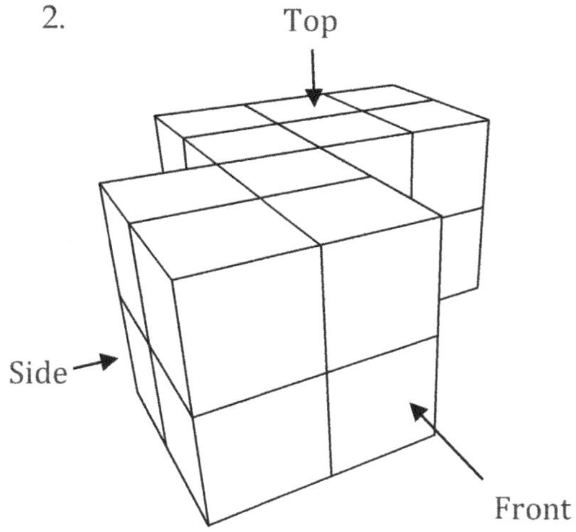

Front View | Top View | Side View

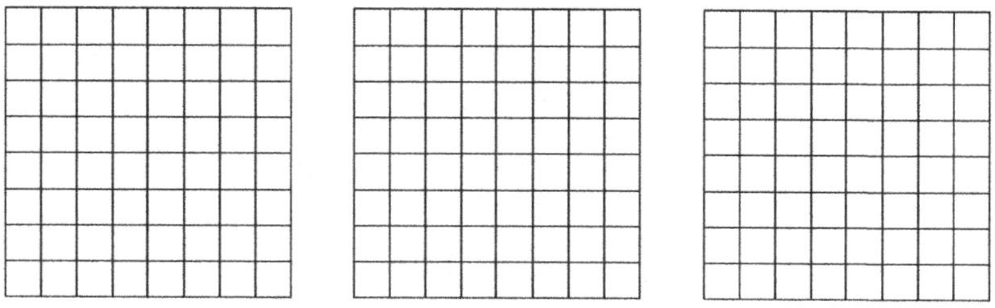

Total number of cubes in the pile: 20

3.

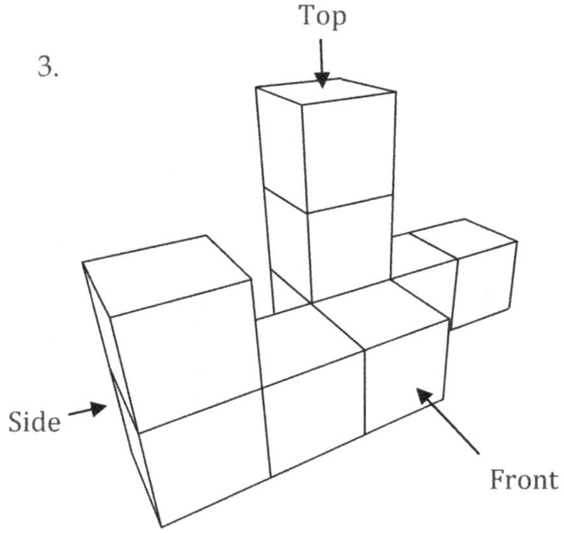

Front View | Top View | Side View

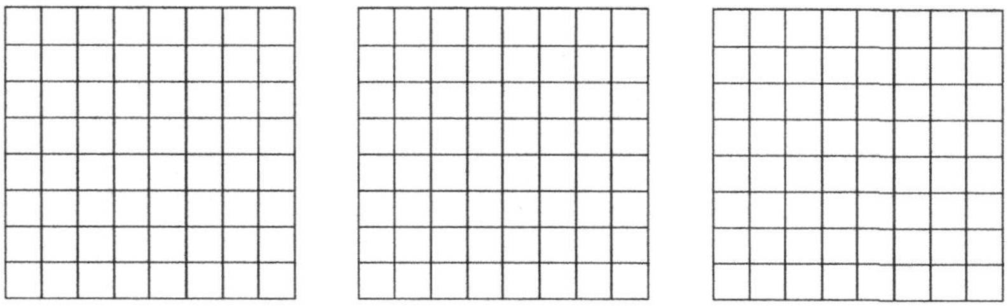

Total number of cubes in the pile: 9

4.

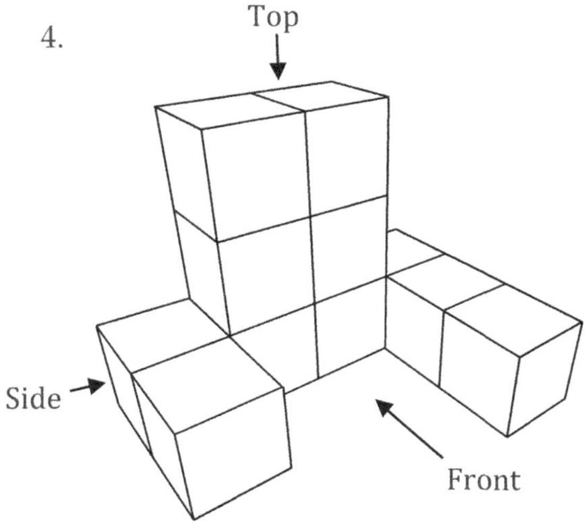

Front View | Top View | Side View

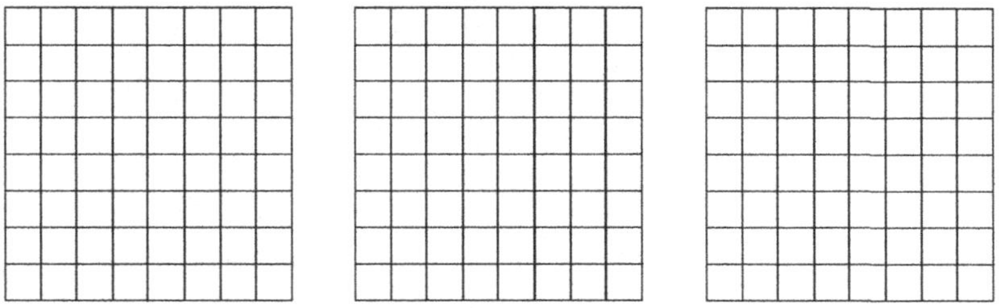

Total number of cubes in the pile: 11

5.

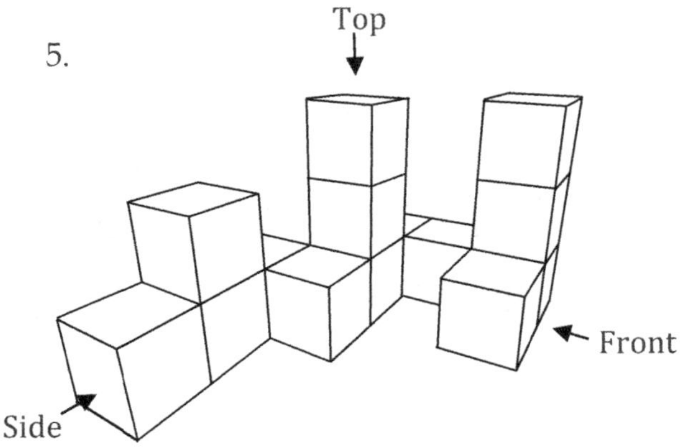

Front View | Top View | Side View

Total number of cubes in the pile: 14

6.

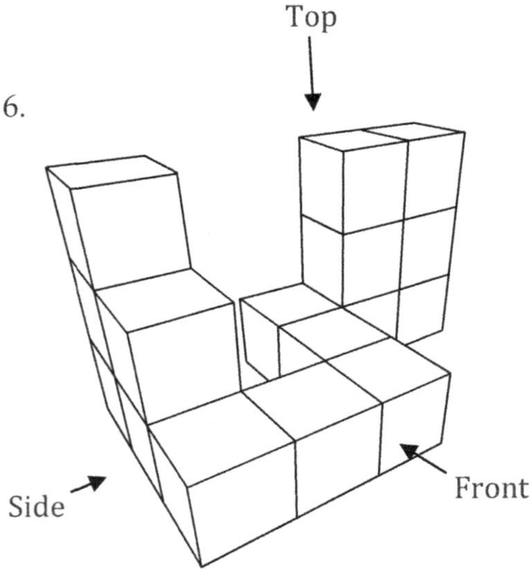

Front View | Top View | Side View

Total number of cubes in the pile: 16

7.

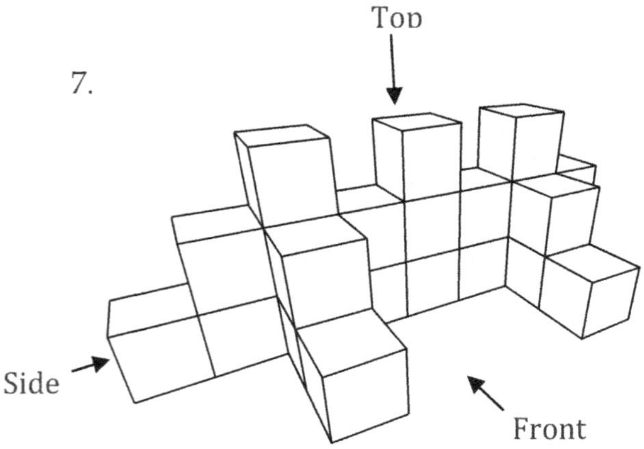

Front View | Top View | Side View

Total number of cubes in the pile: 24

8.

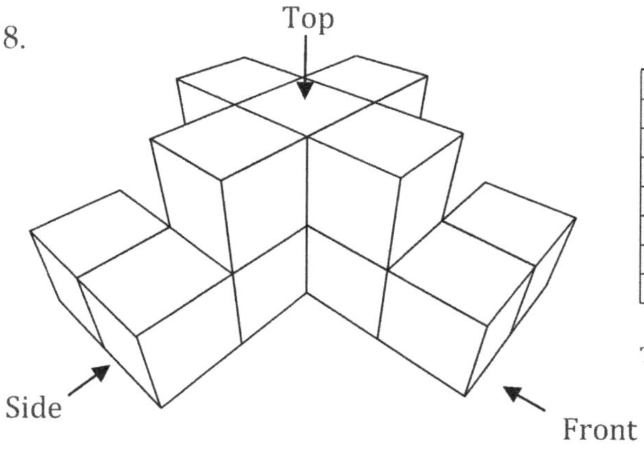

Front View | Top View | Side View

Total number of cubes in the pile: 14

9.

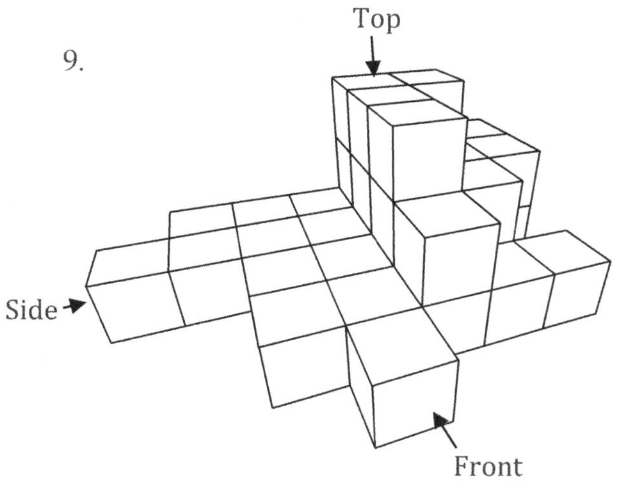

Front View | Top View | Side View

Total number of cubes in the pile: 36

10.

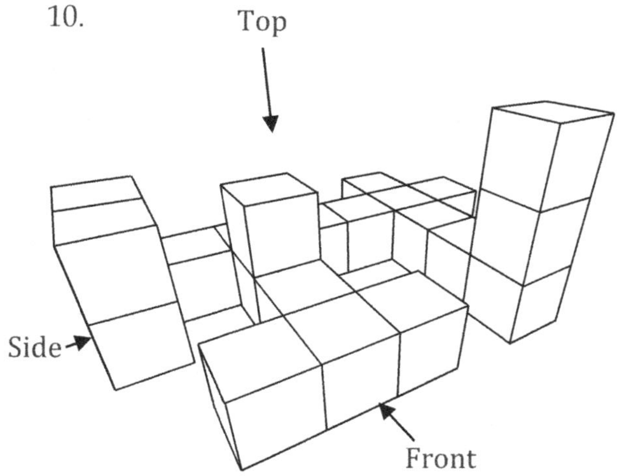

Front View | Top View | Side View

Total number of cubes in the pile: 22

Top View, Front View, and Side View

Part 3

For each question, choose one 3D-image that matches all the specified views.

1.

Top View

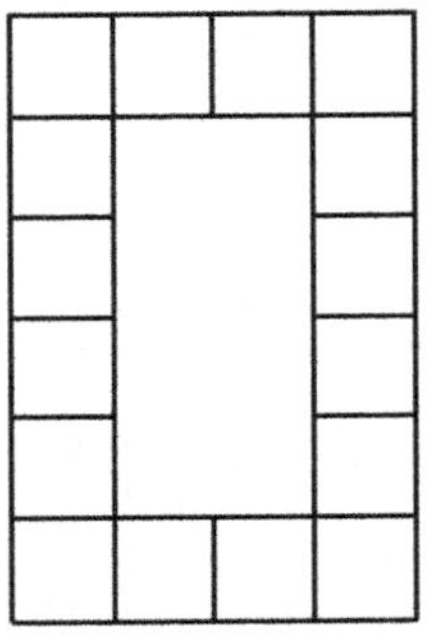

Front View

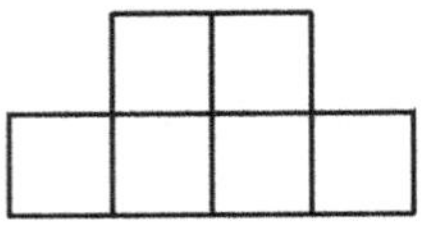

Side View

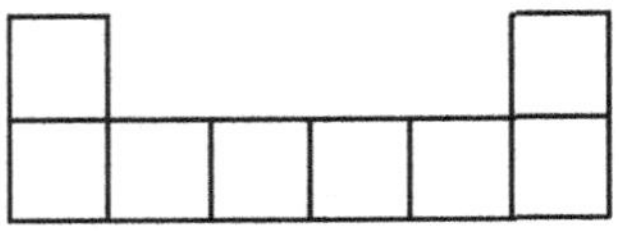

a.

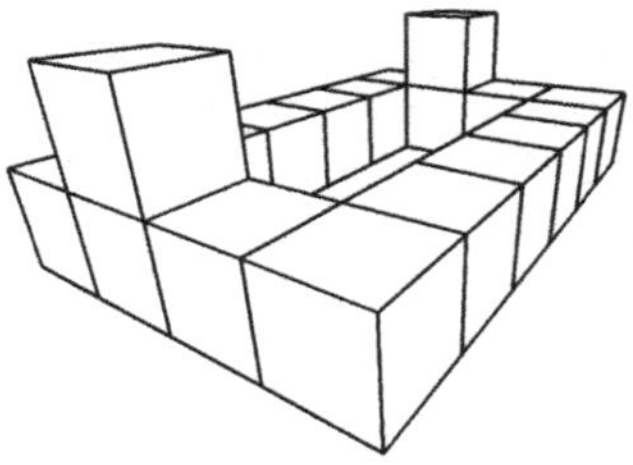

b.

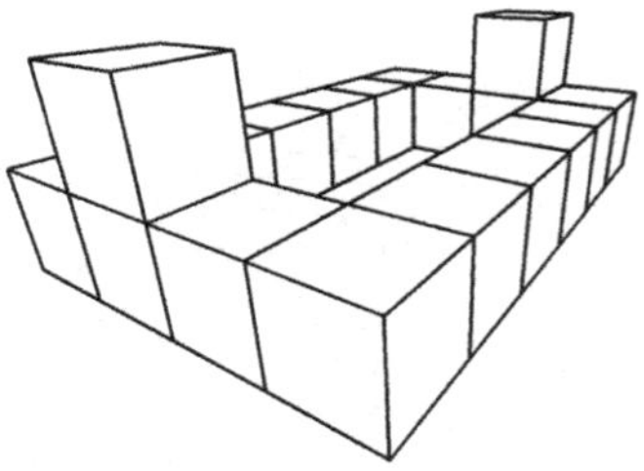

c.

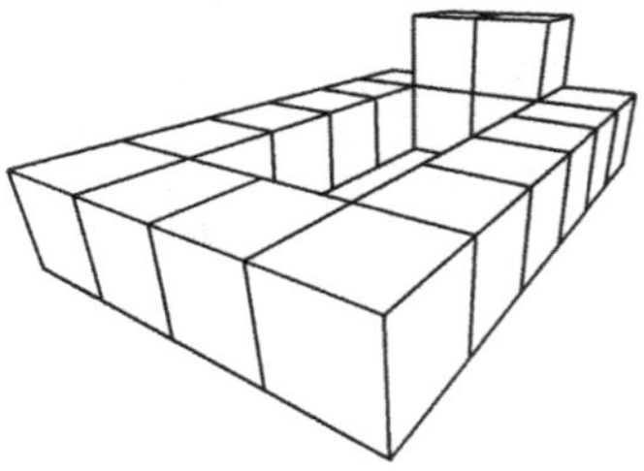

d.

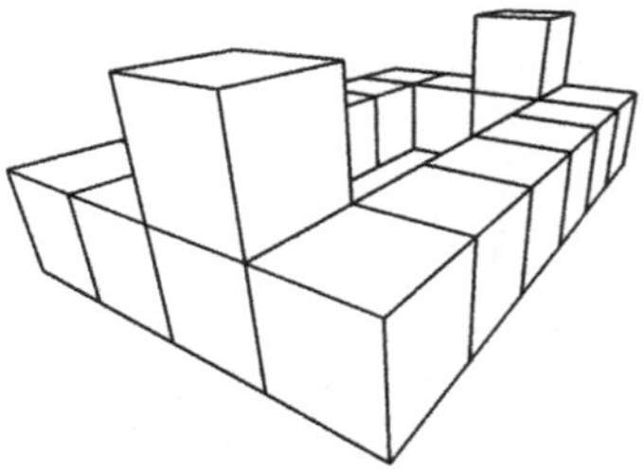

e.

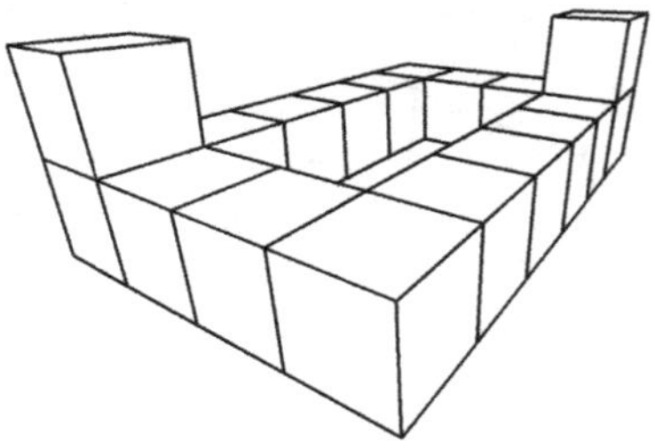

2.

Top View

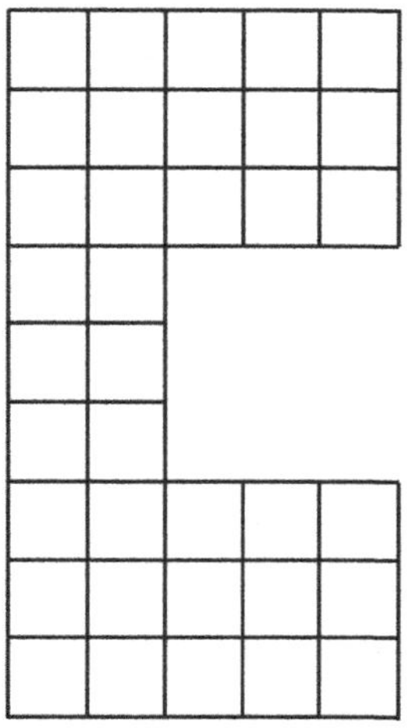

Front View

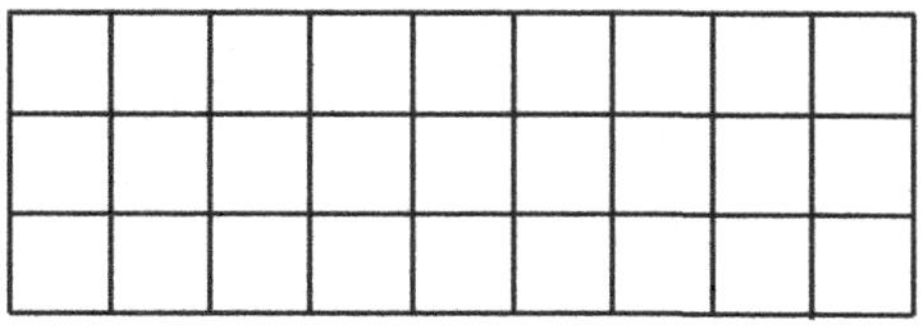

Side View

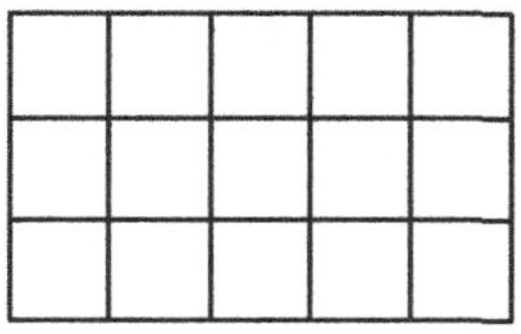

a.

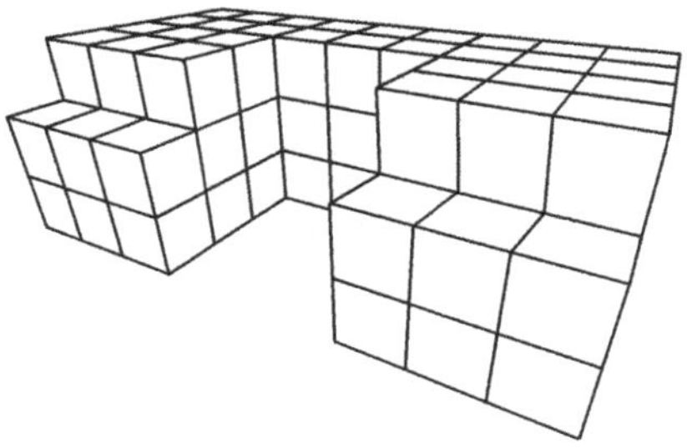

b.

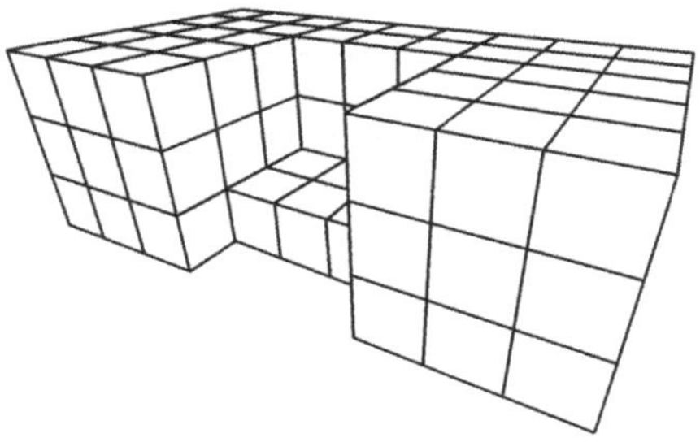

c.

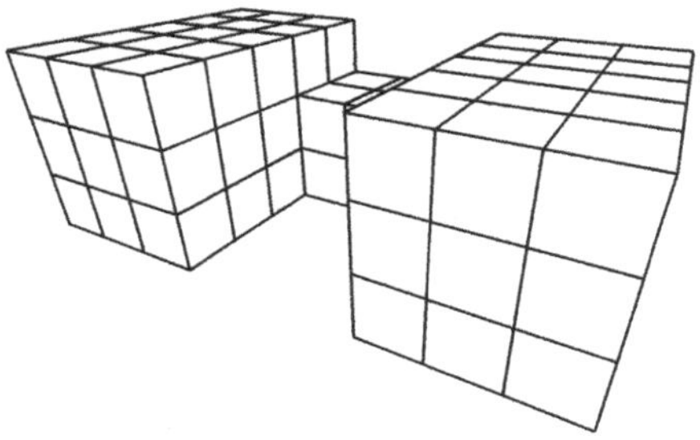

d.

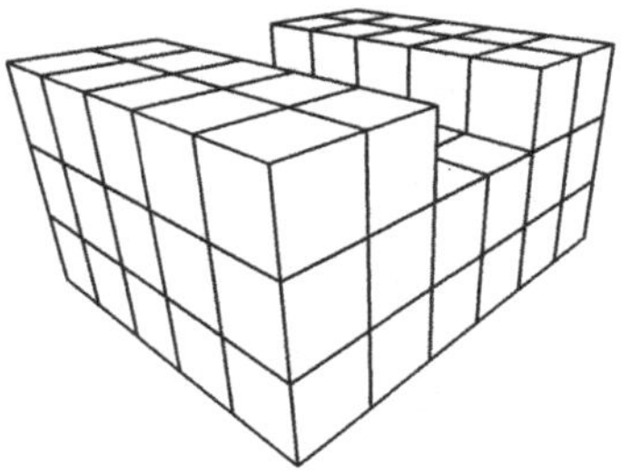

e.

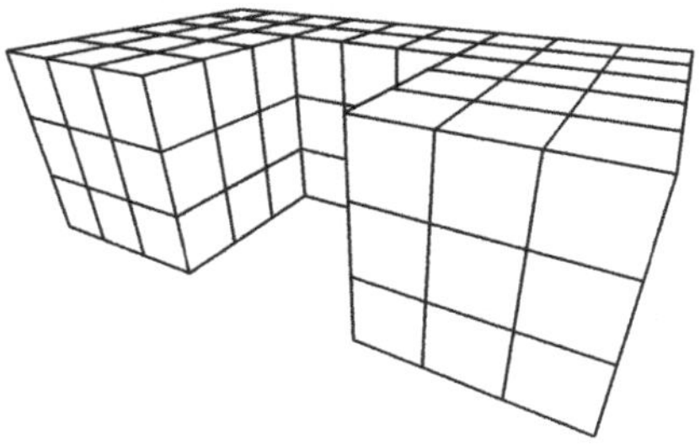

3.

Top View

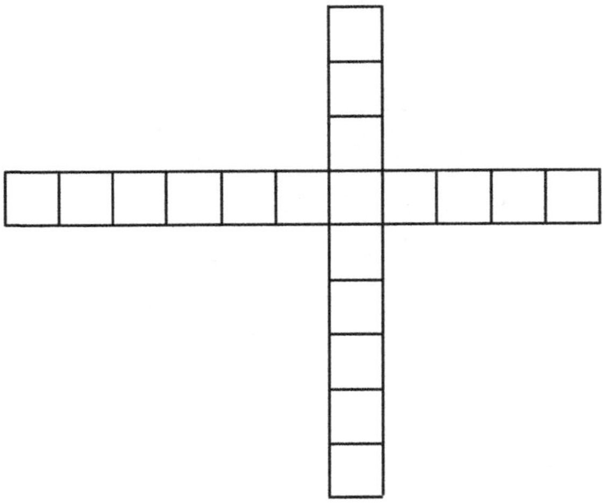

Front View

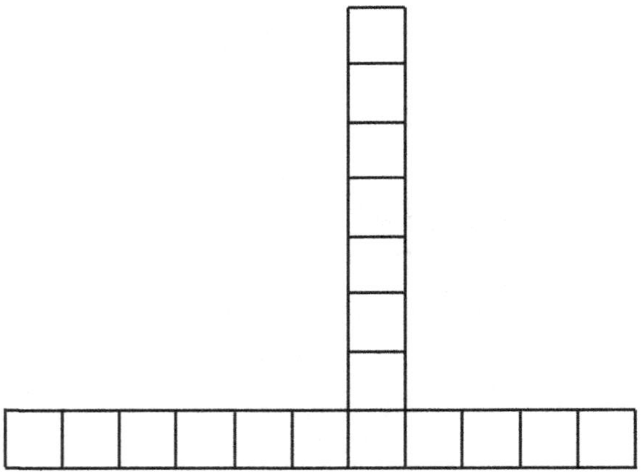

a.

b.

c.

d.

e.

4.

Top View

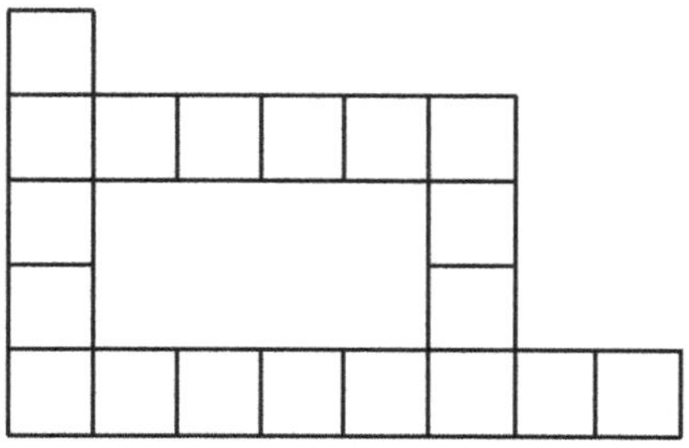

Front View

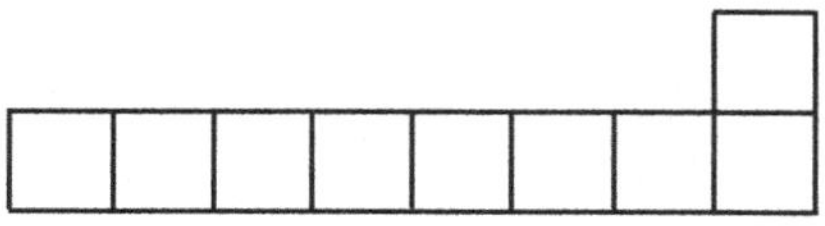

Side View

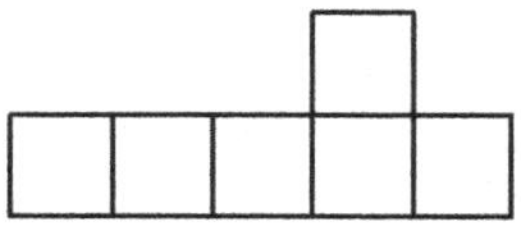

a.

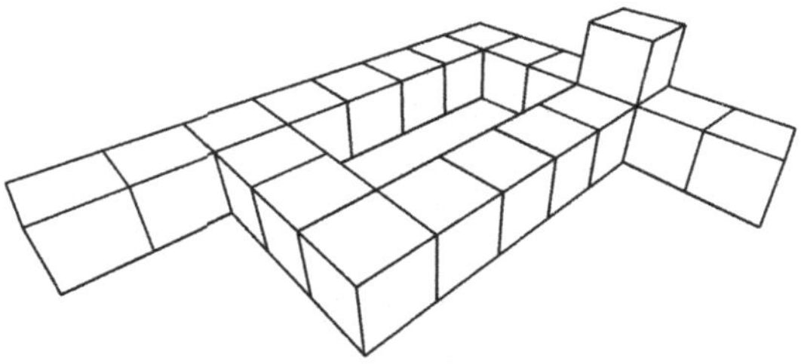

b.

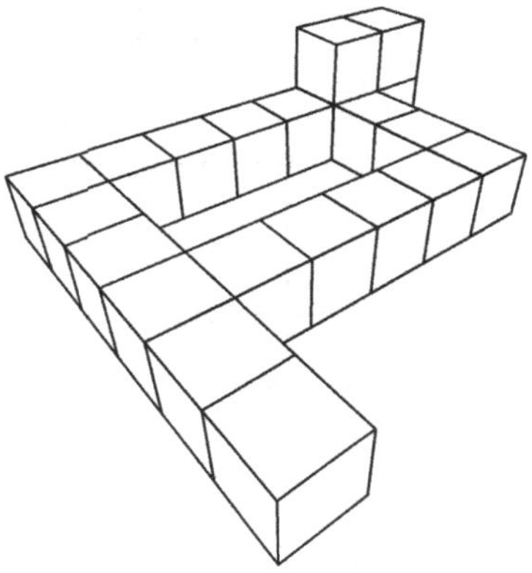

c.

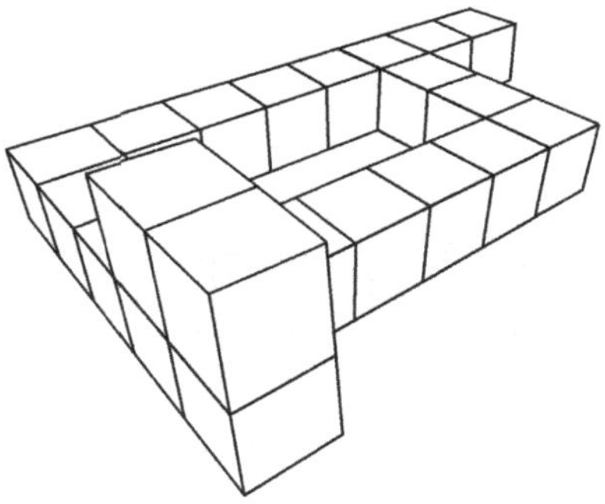

d.

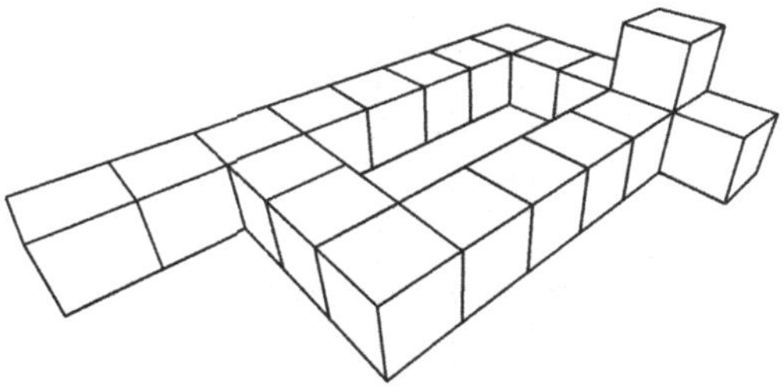

e.

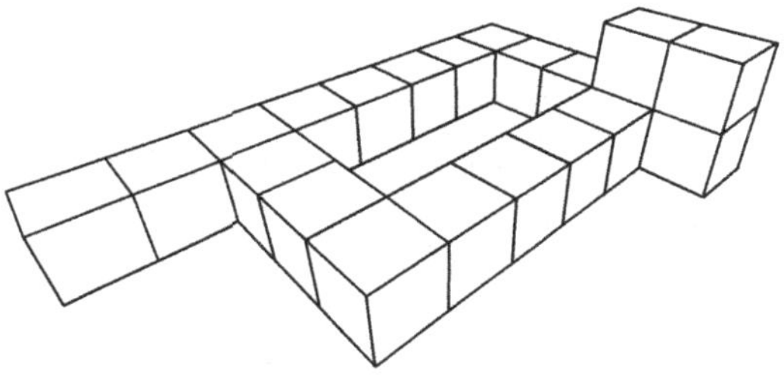

5.

Top View

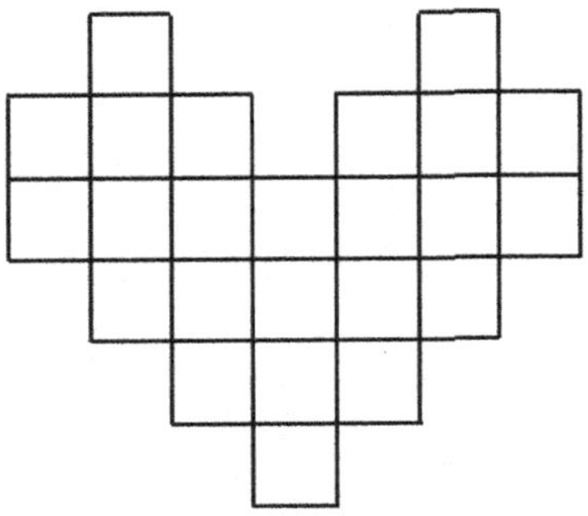

Front View

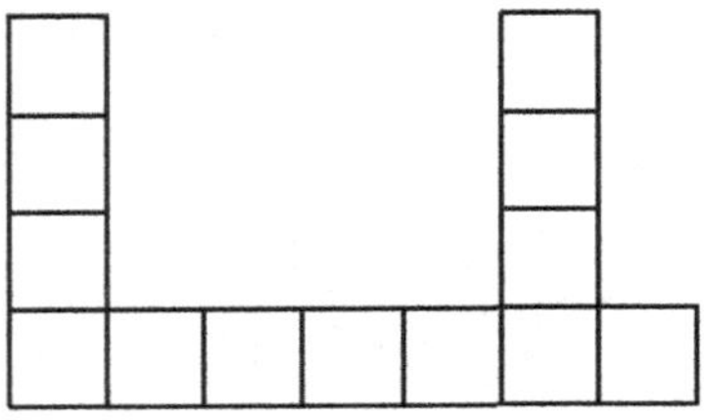

Side View

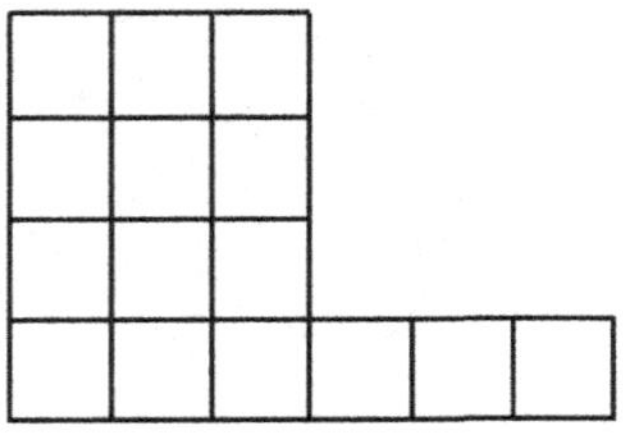

a.

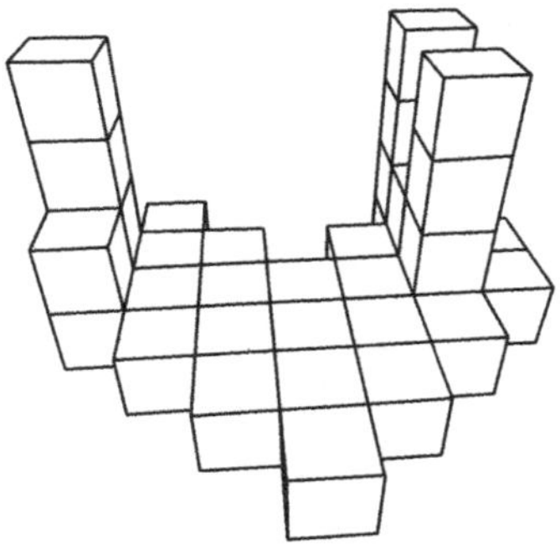

b.

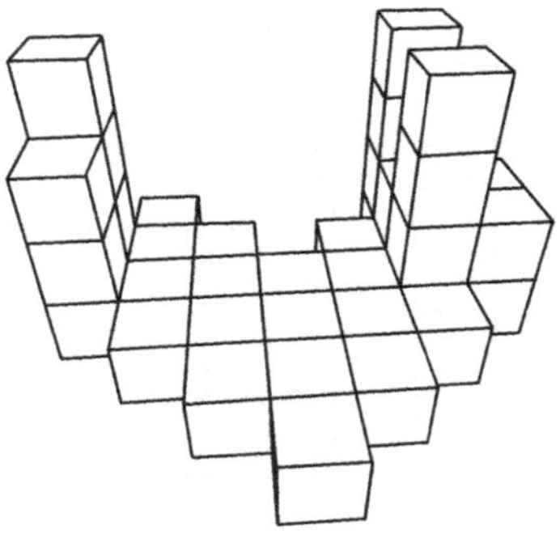

c.

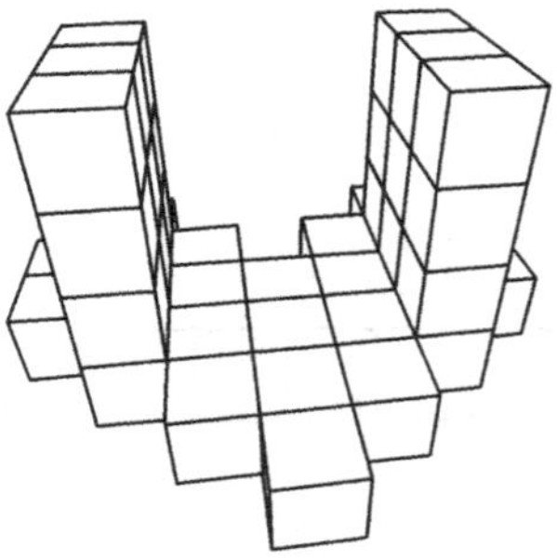

d.

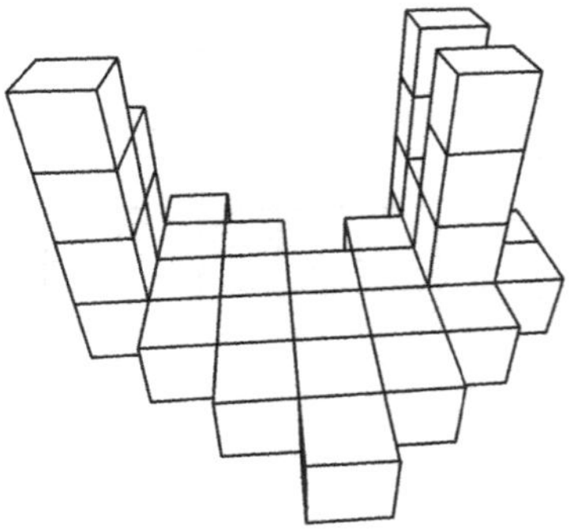

e.

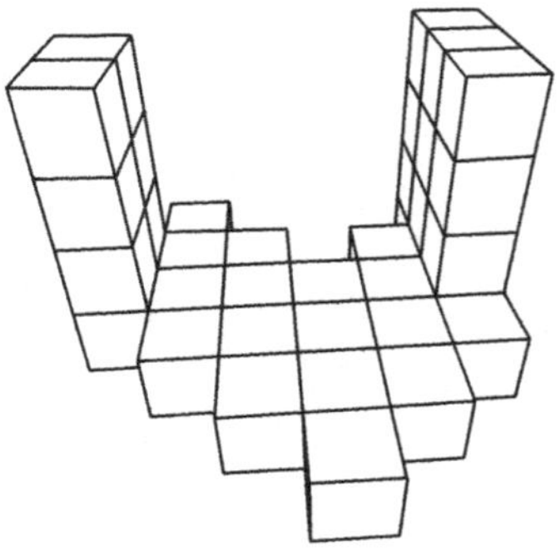

6.

Top View

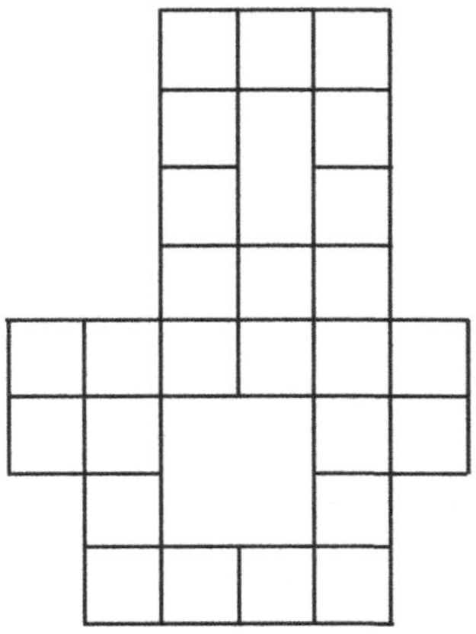

Front View

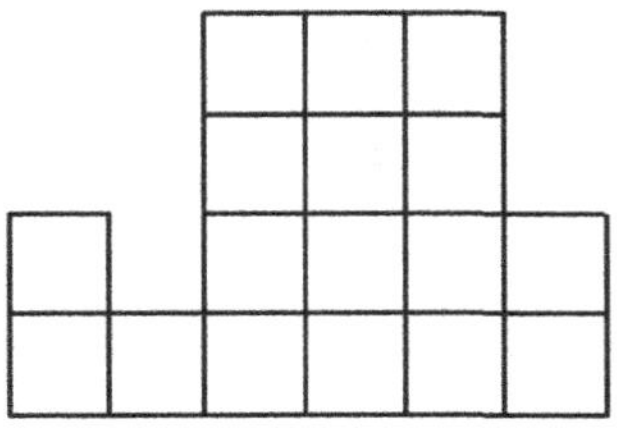

Side View

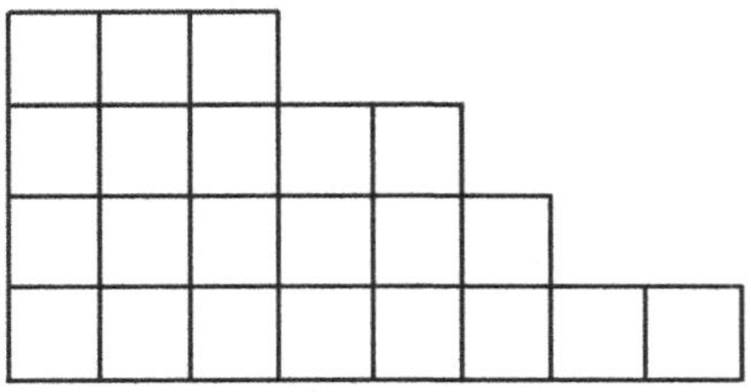

a.

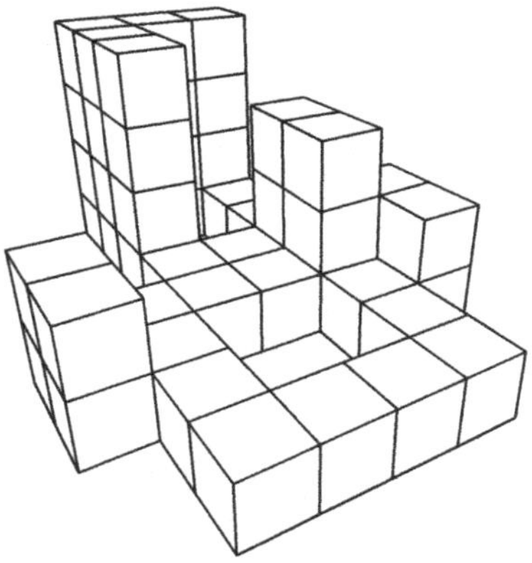

b.

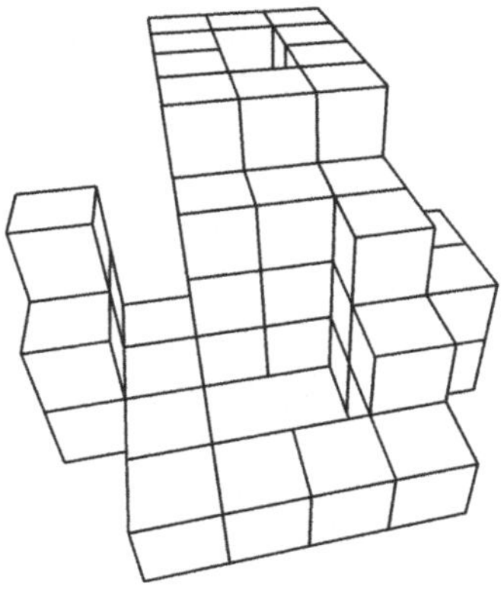

c.

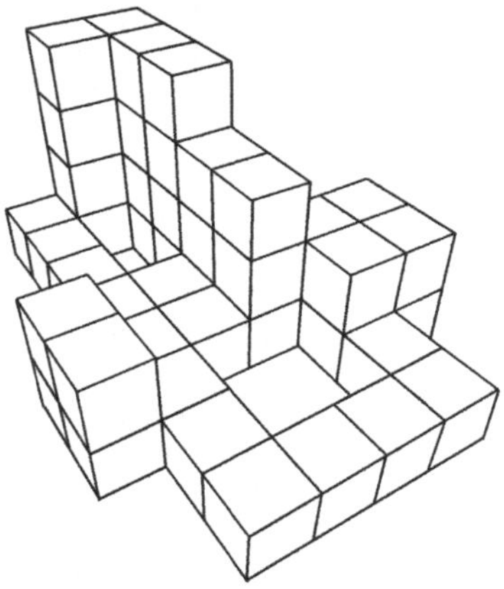

d.

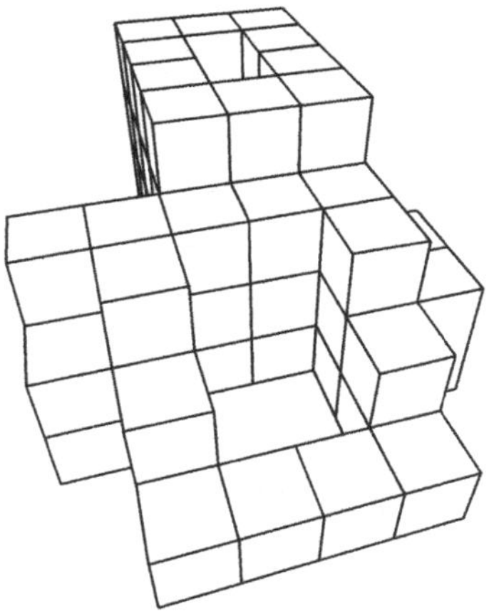

7.

Top View

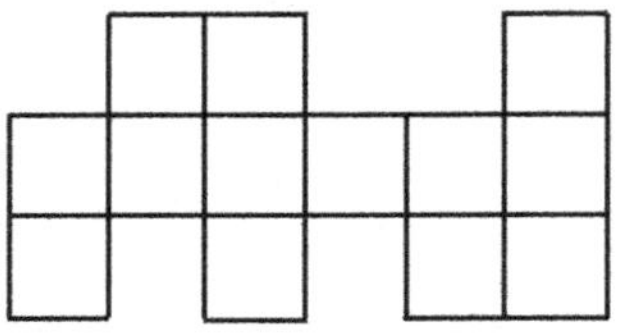

Front View

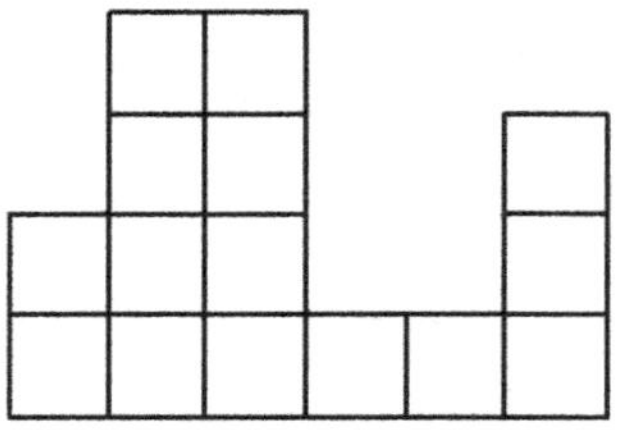

Side View

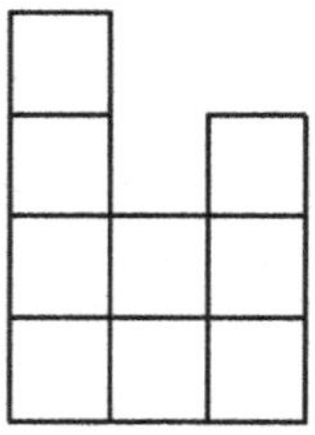

a.

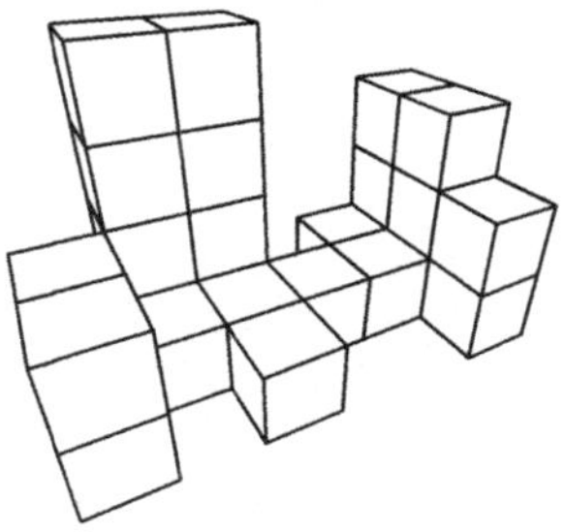

b.

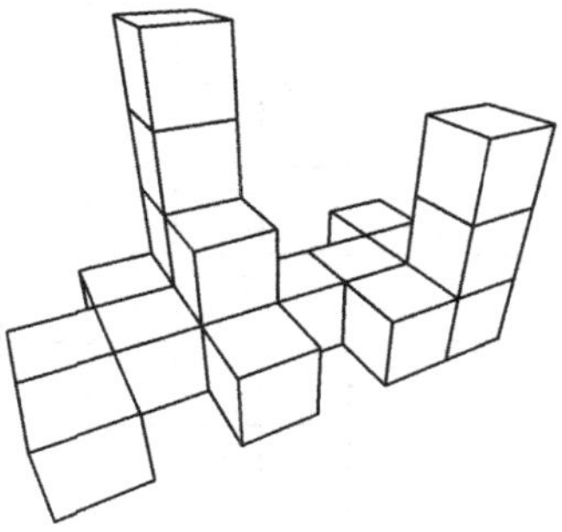

c.

d.

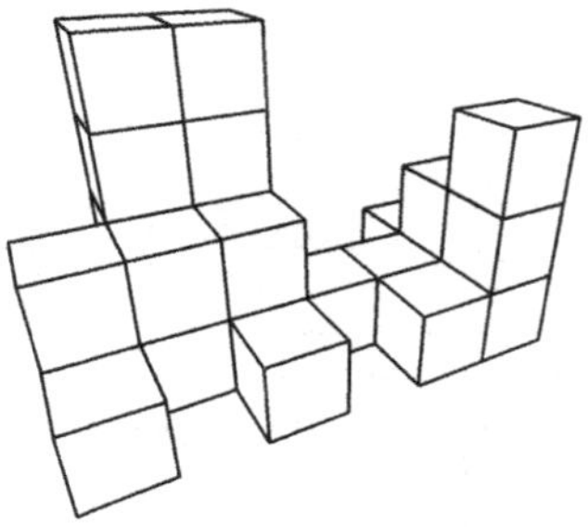

e.

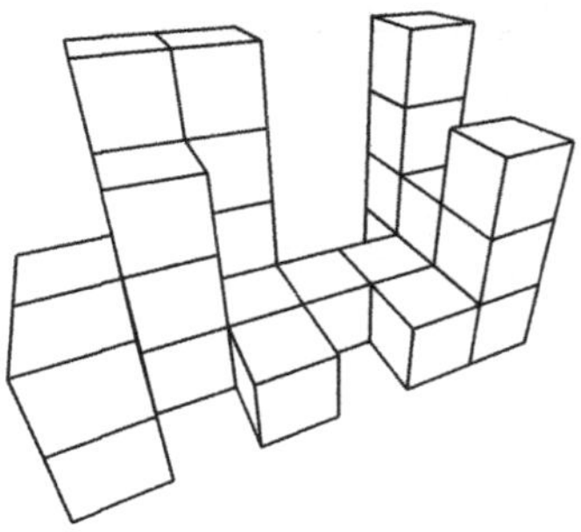

8.

Top View

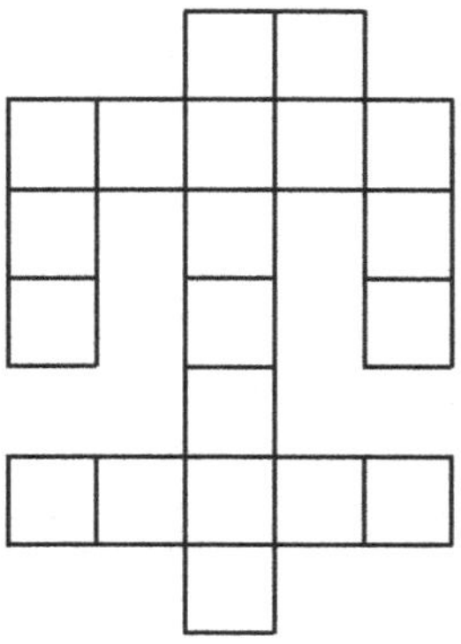

Front View

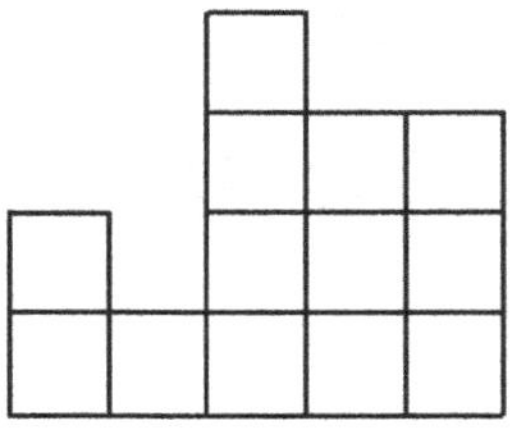

Side View

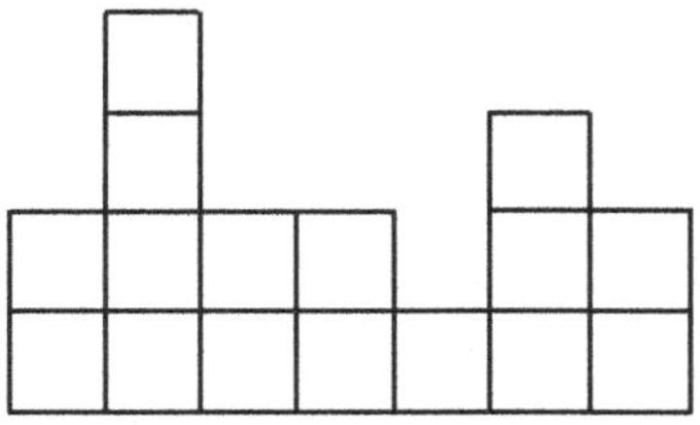

a.

b.

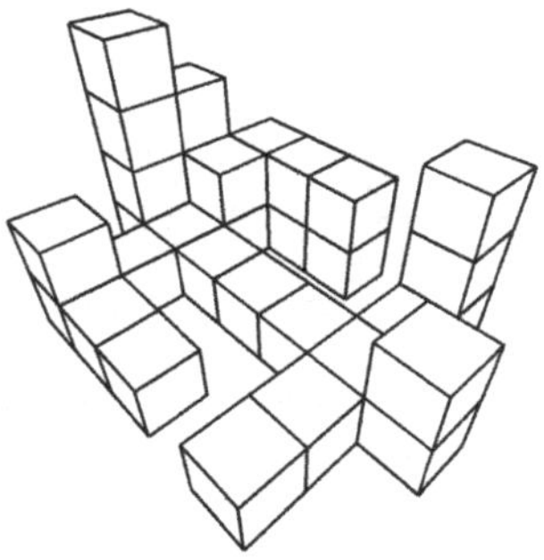

c.

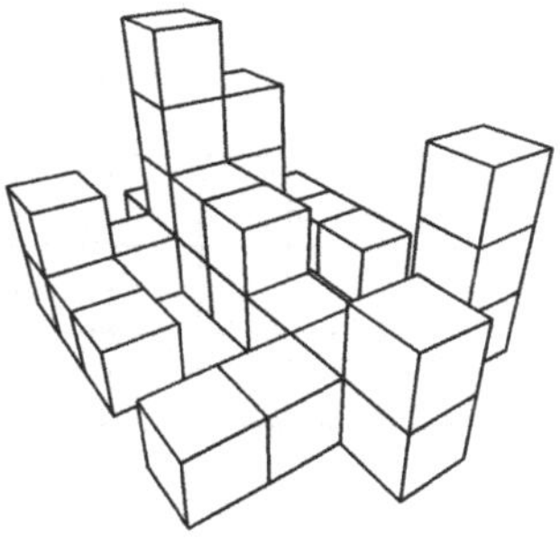

d.

e.

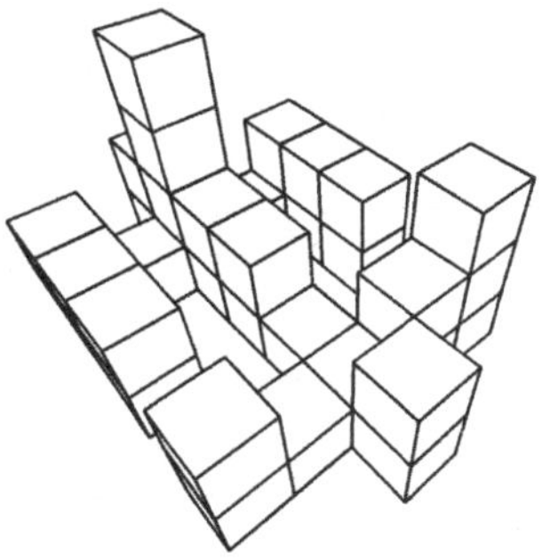

9.

Top View

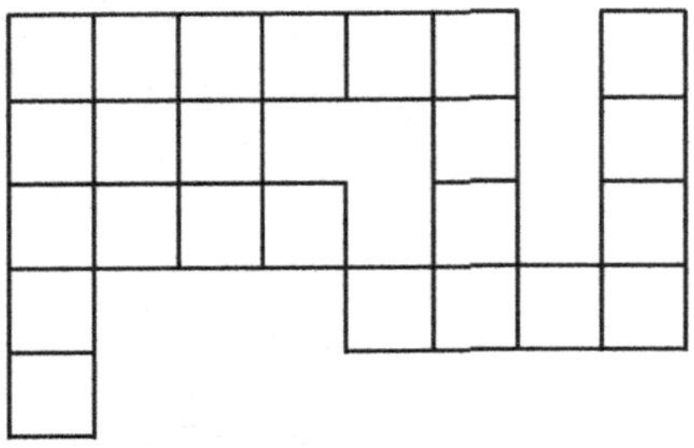

Front View

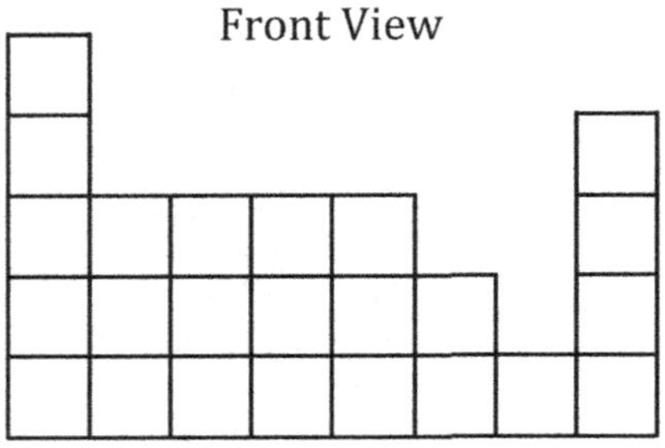

Side View

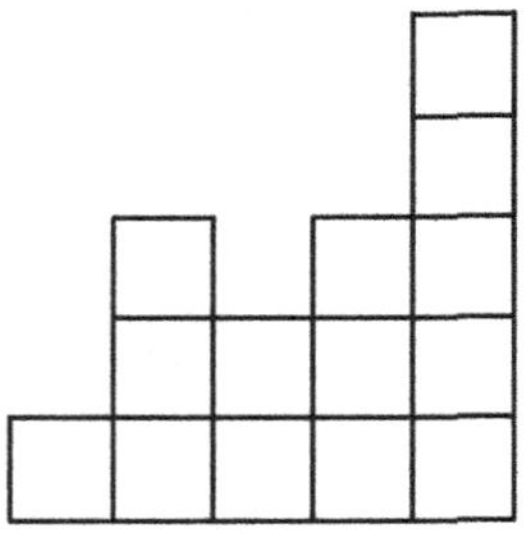

a.

b.

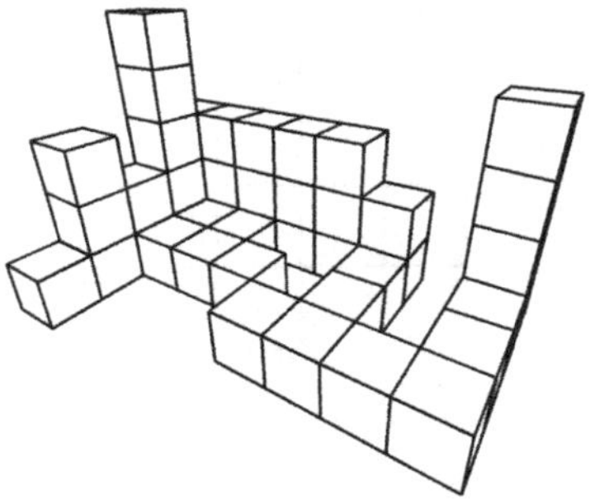

c.

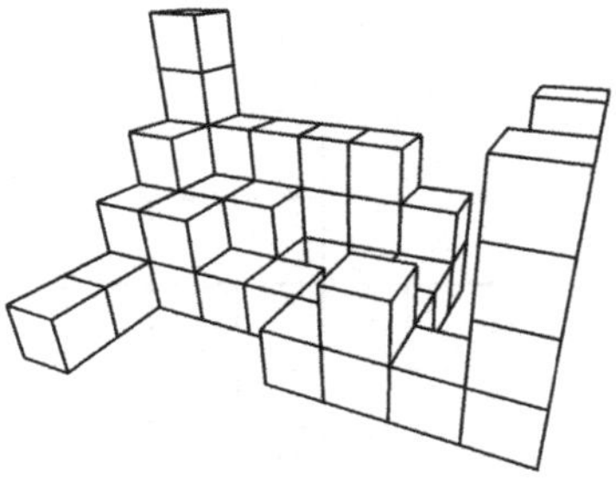

d.

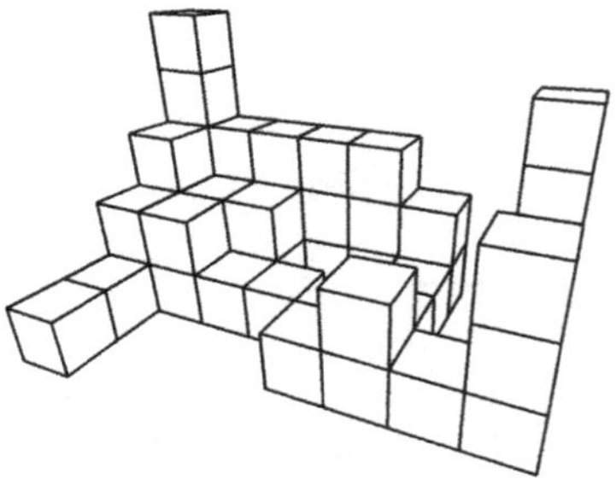

e.

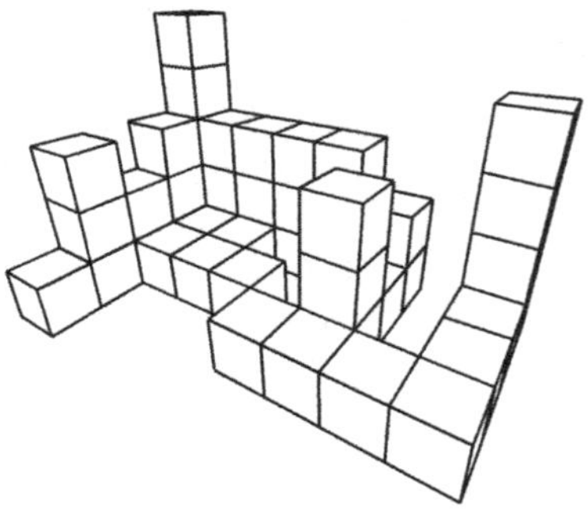

10.

Top View

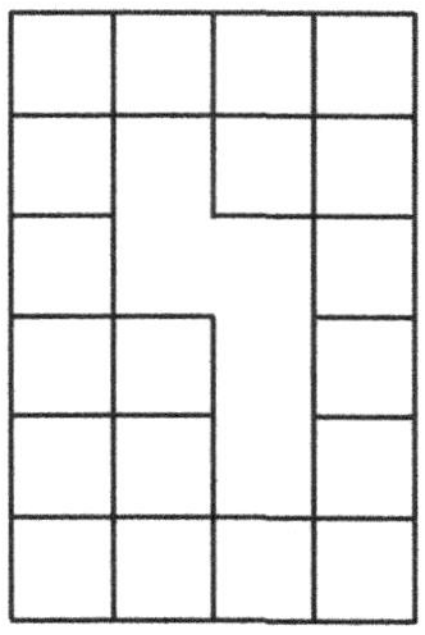

Front View

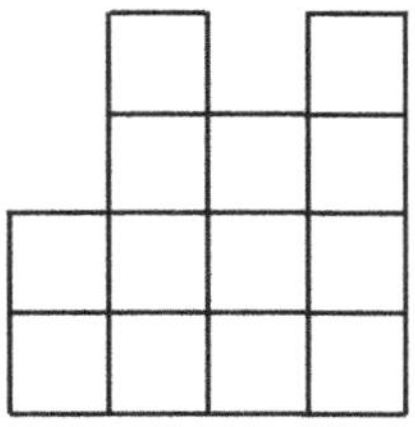

Side View

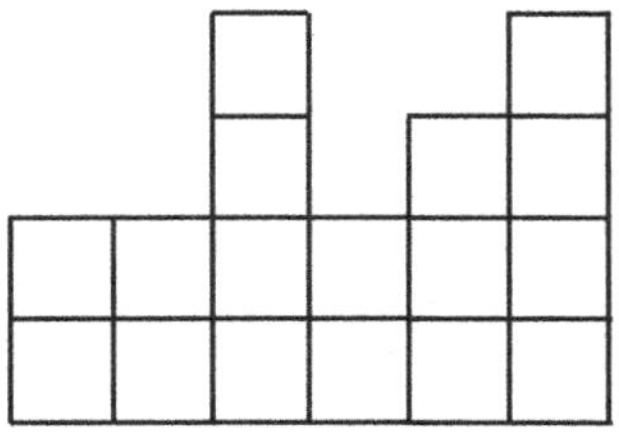

a.

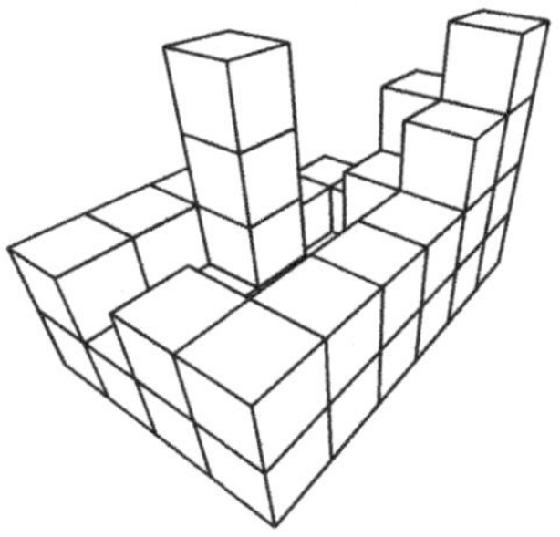

b.

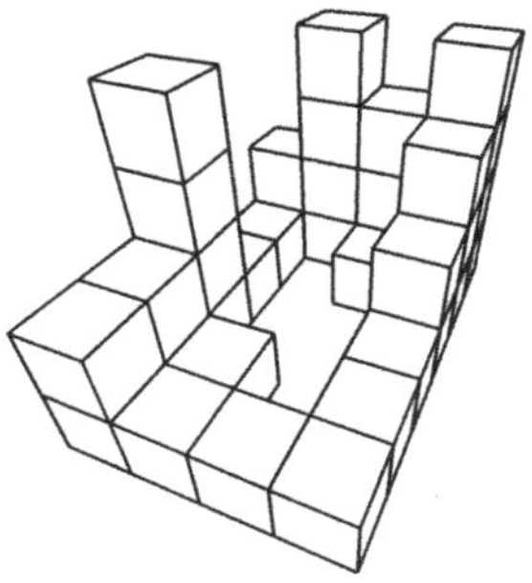

c.

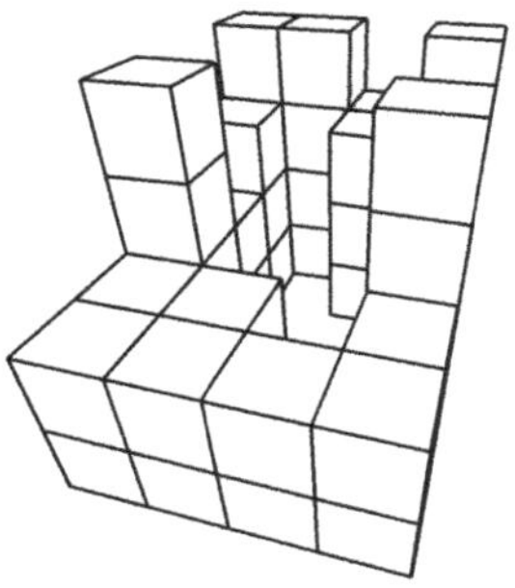

d.

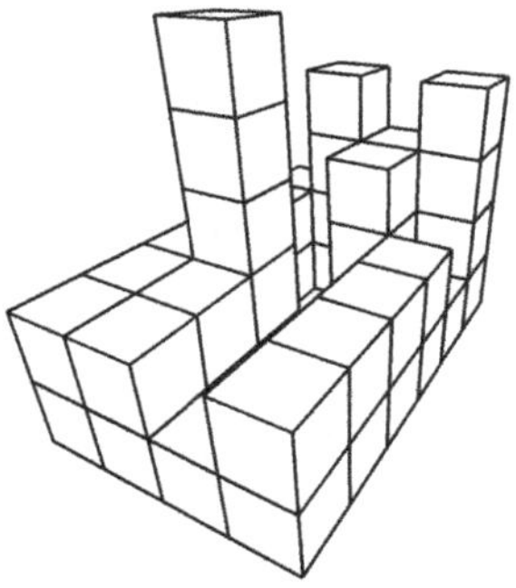

e.

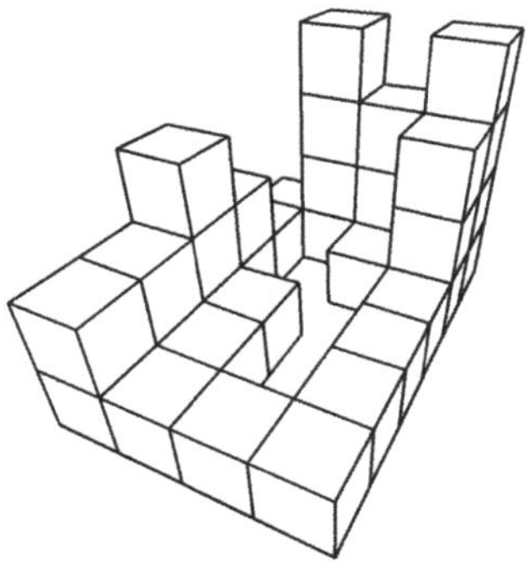

Next Page

Unit Test # 1

Estimated Duration: 1 hour

Unit Test

Top View, Front View, and Side View

____
40

Note, for every question, every cube is the same size and shape.
Some of cubes may be hidden from plain view and no cube is floating in the air.
If a cube is not laying on the ground, there must be another cube underneath holding it in place.

Multiple Choice

1. Circle the top view for the given structure.

Top

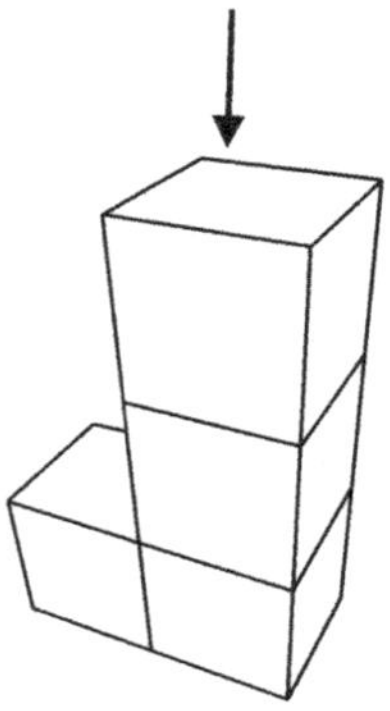

a. 

b. 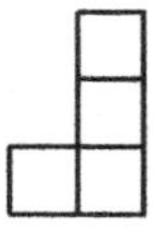

c. 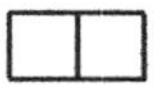

d. 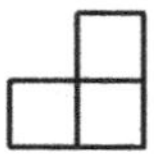

2. Circle the front view for the given structure.

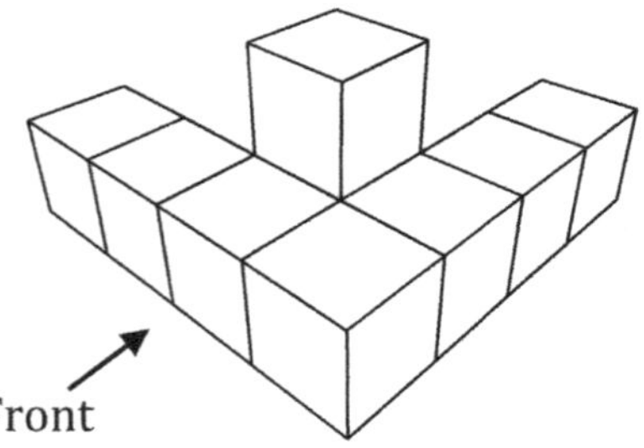

Front

a. 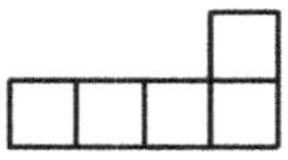

b. 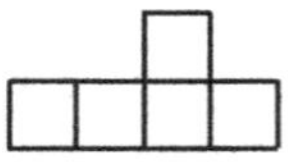

c. 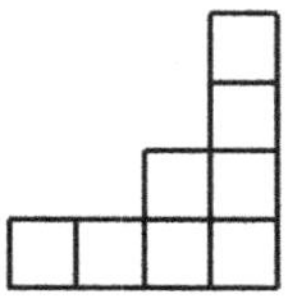

d. 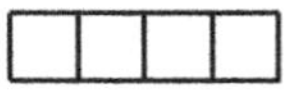

3. Circle the side view for the given structure.

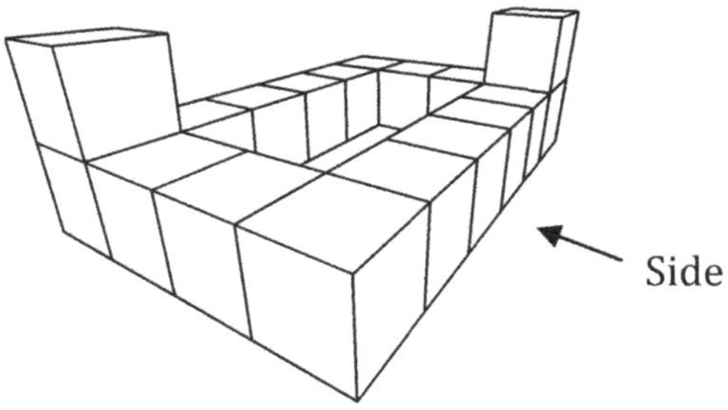

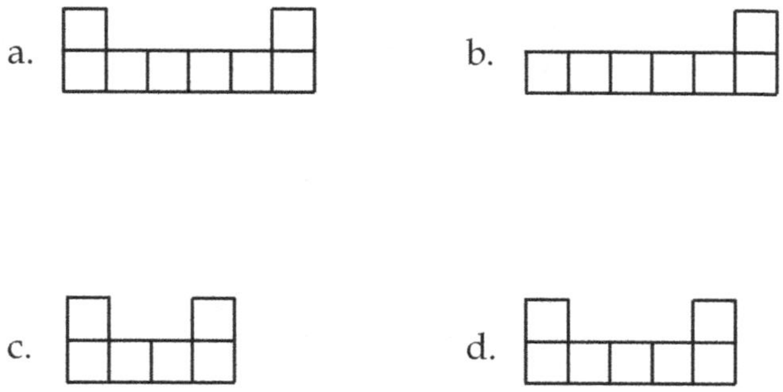

4. Circle the side view for the given structure.

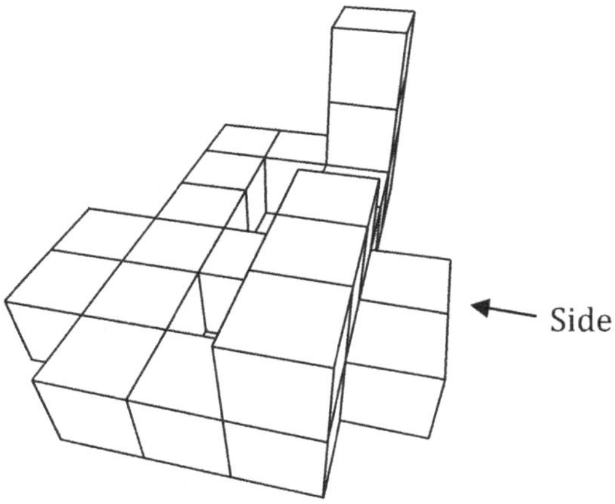

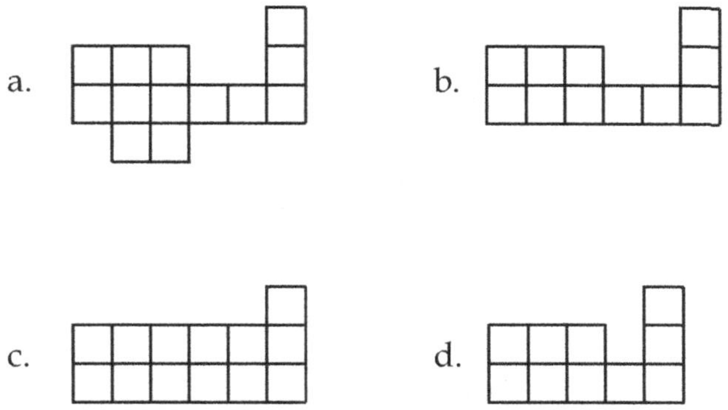

5. Circle the front view for the given structure.

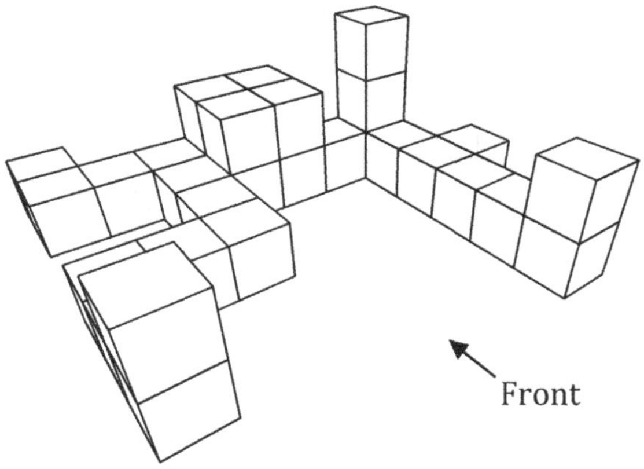

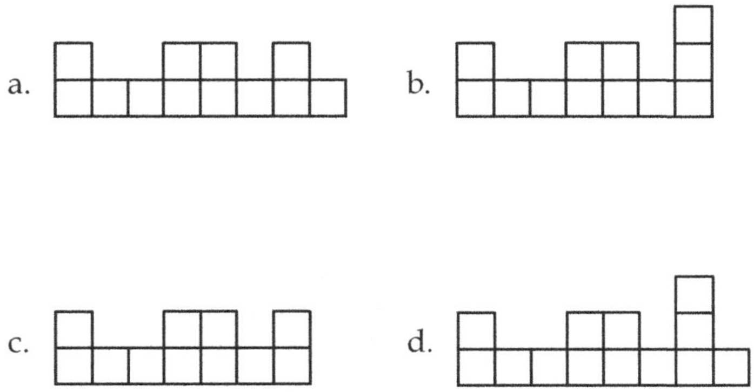

6. Circle the top view for the given structure.

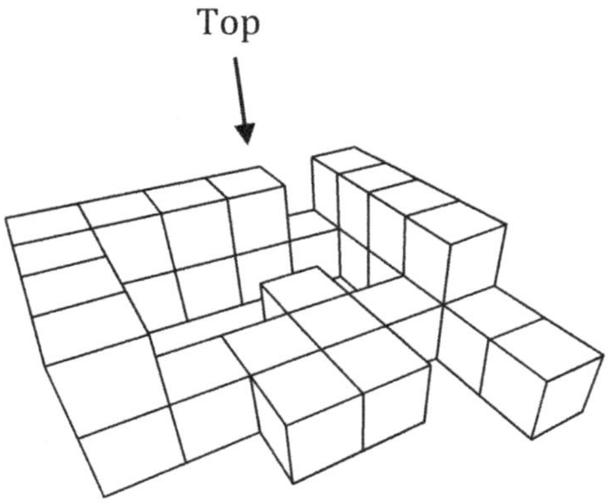

a. 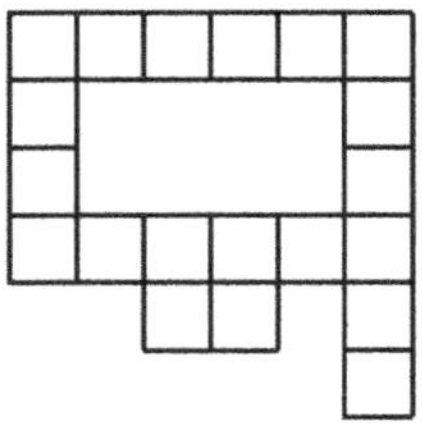

b. 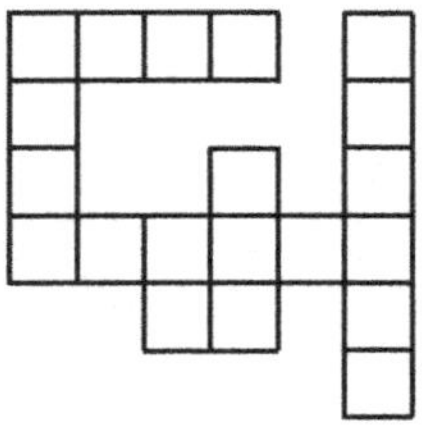

c. 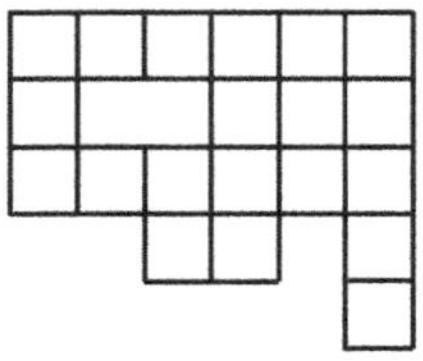

d. 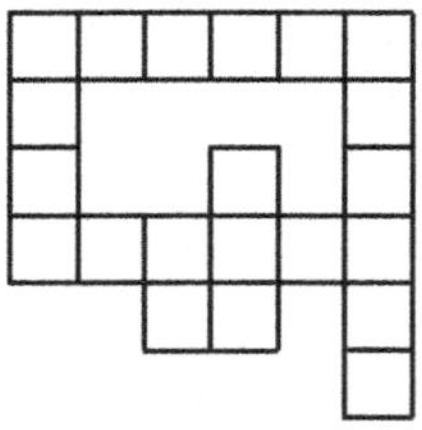

7. Sebastian is standing in front of the pile of cubes as shown below. Circle the option that Sebastian sees.

a. 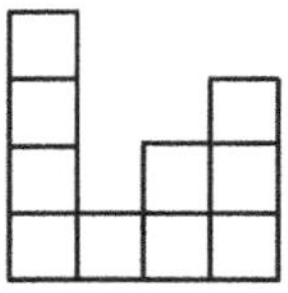

b. 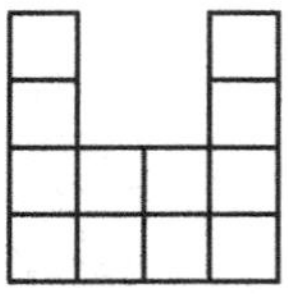

c. 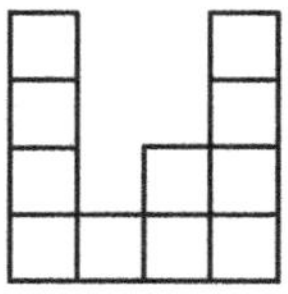

d. 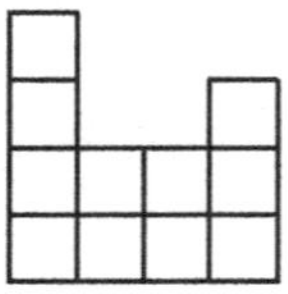

8. A plane is flying over the cube structure as shown below. Circle the plane's perspective of the structure.

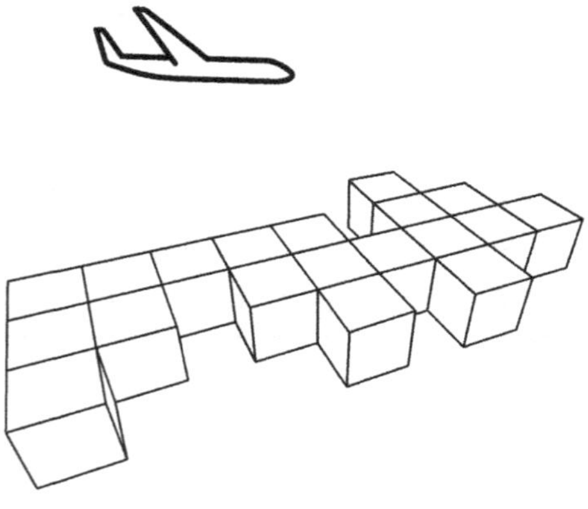

a. 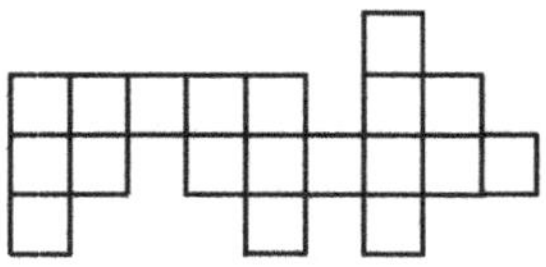

b. 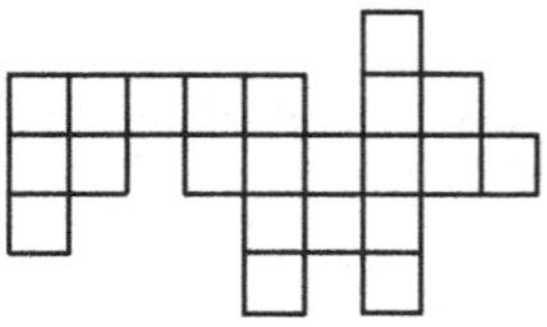

c. 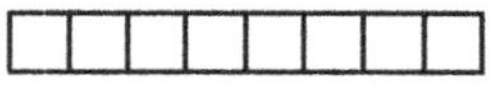

d. 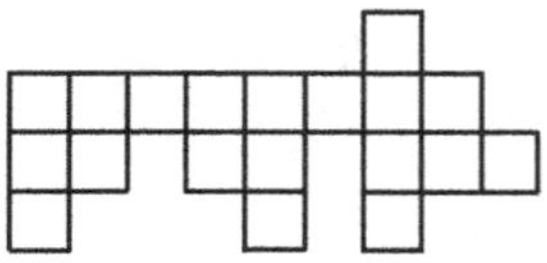

9. Circle the option that best fills in the blank.

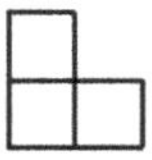 is the _____________ view for the given shape:

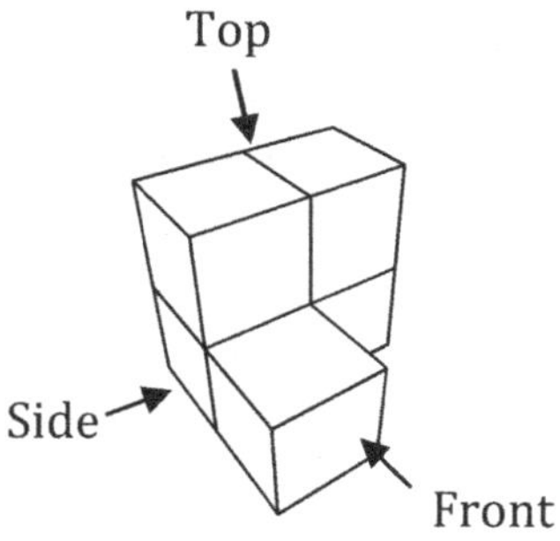

a. top
b. front
c. side
d. top & side
e. top & front
f. front & side

10. Circle the option that best fills in the blank.

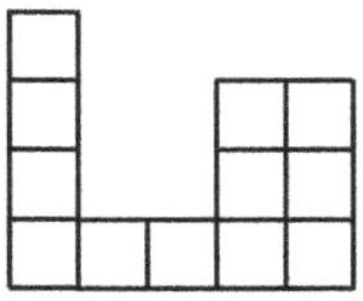 is the _____________ view for the given shape:

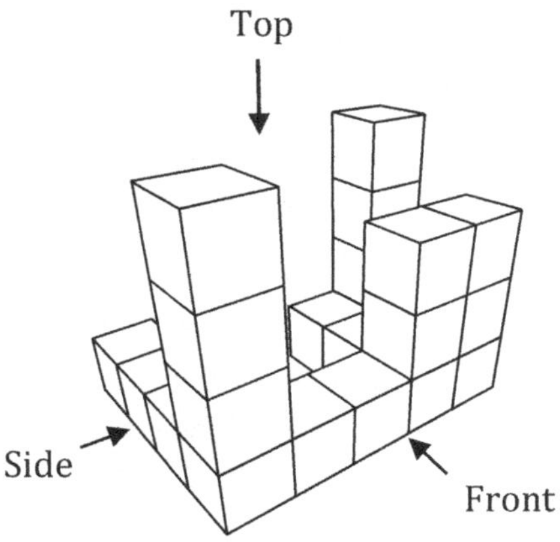

a. top
b. front
c. side
d. top & side
e. top & front
f. front & side

11. Circle the option that best fills in the blank.

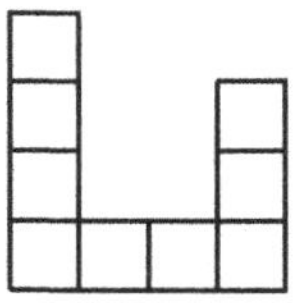 is the _____________ view for the given shape:

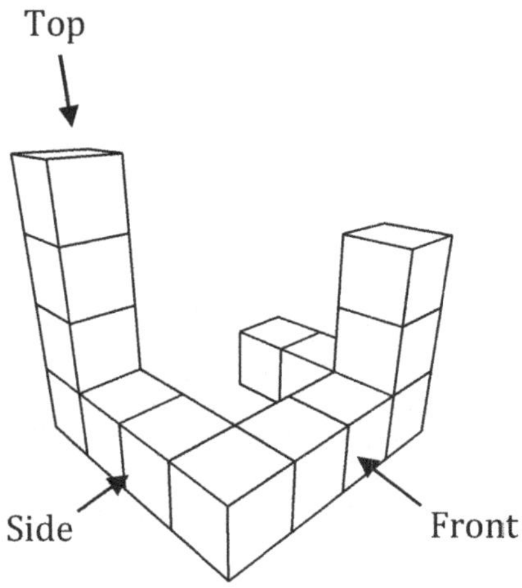

a. top
b. front
c. side
d. top & side
e. top & front
f. front & side
g. top, side, & front
f. none of the above

12. Circle the one 3D-image that matches all the specified views.

Top

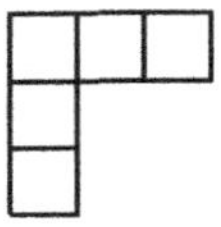

Front

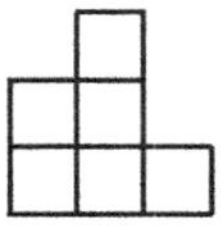

Side

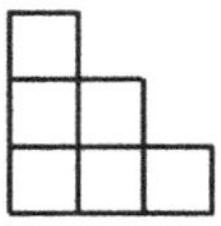

a.

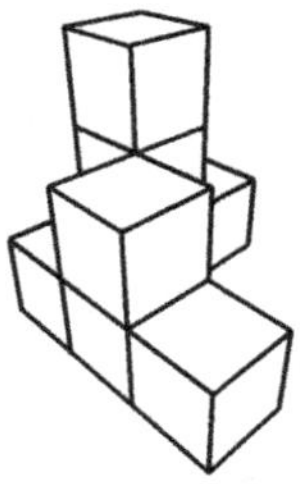

b.

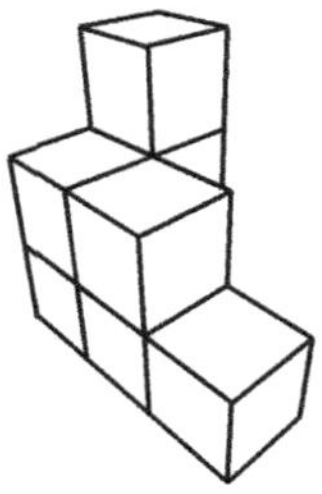

c.

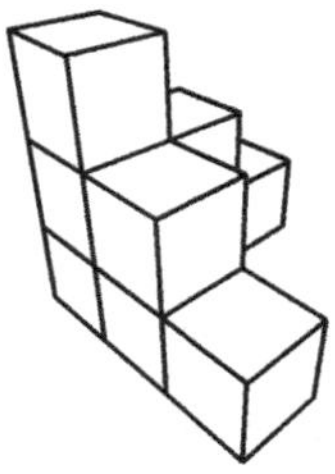

d.

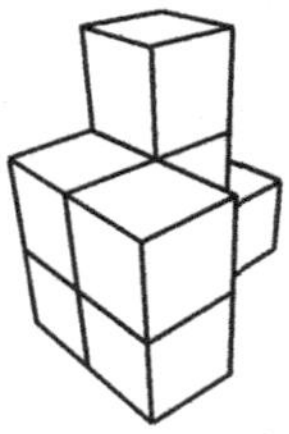

13. Circle the one 3D-image that matches all the specified views.

Top

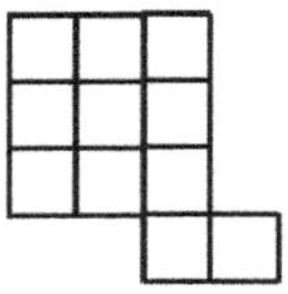

Front

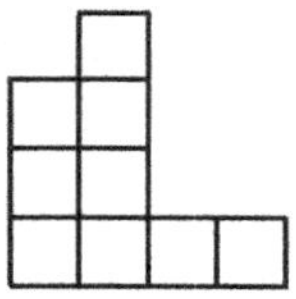

Side

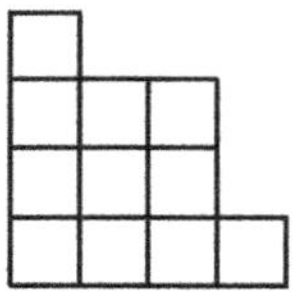

a.

b.

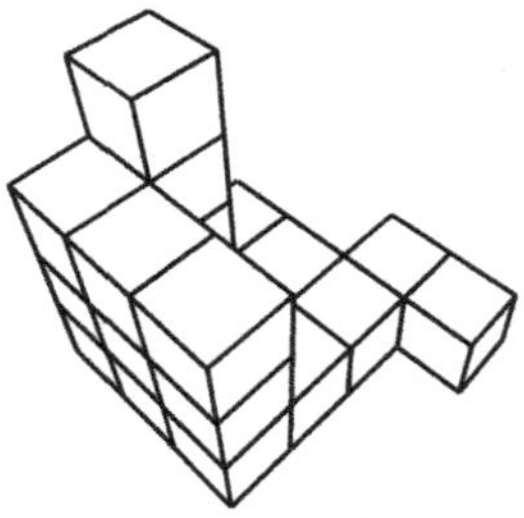

c.

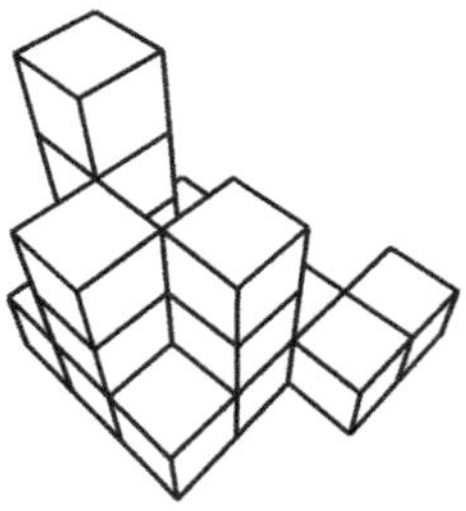

d.

14. Circle the one 3D-image that matches all the specified views.

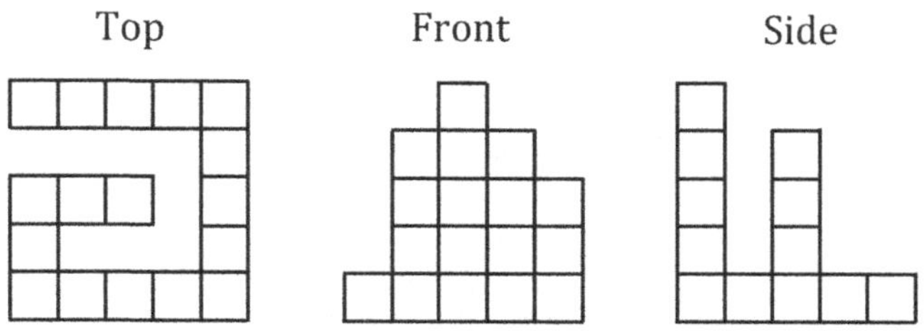

a.

b.
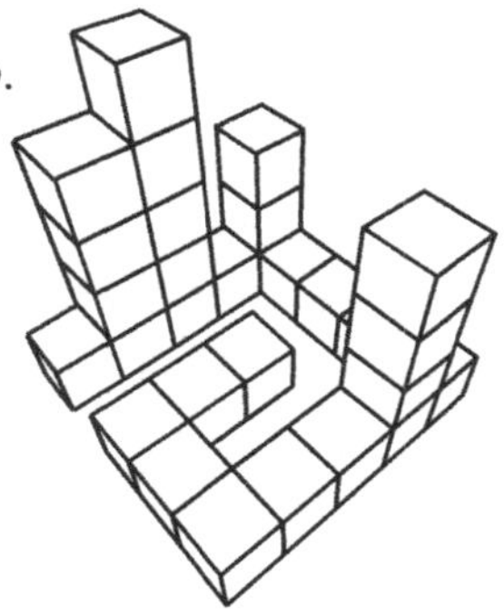

c.

d.

15. Isabella is looking at a structure from the side and Sebastian is looking at the same structure from the front as shown below. The top view is also provided.

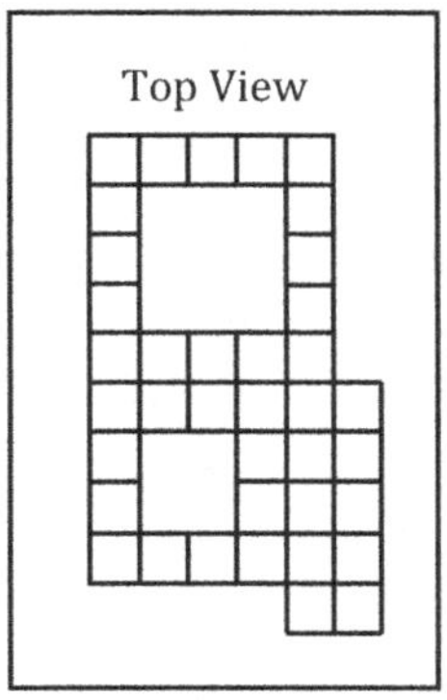

Assuming that no cube is floating in the air, that is, if the cube is not on the ground there must be another cube directly underneath holding it in place, how many <u>more</u> cube faces is Isabella seeing than Sebastian?

a. 39 b. 27 c. 12 d. 11 e. 10

True or False

For questions 16 – 20, refer to the structure given below.

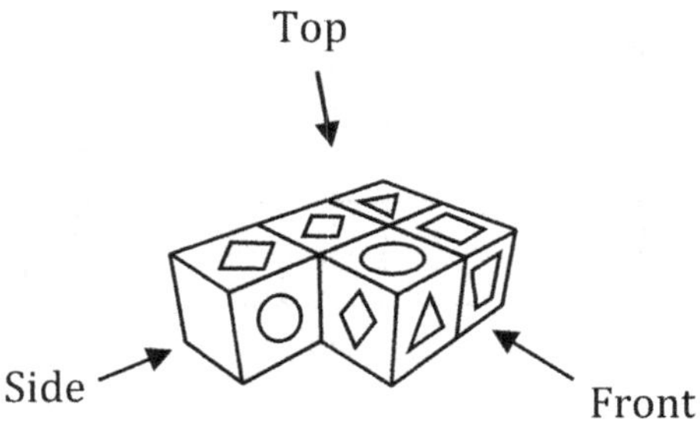

Circle the correct option.

16. 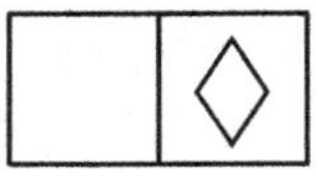 is the side view True or False

17.  is the front view True or False

18. 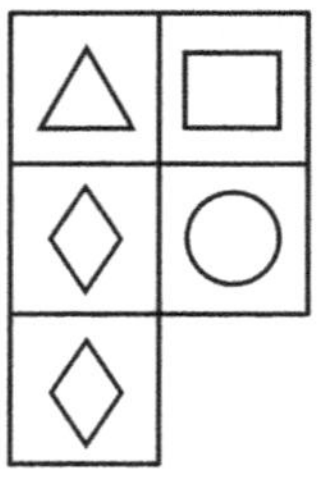 is the front view True or False

19. 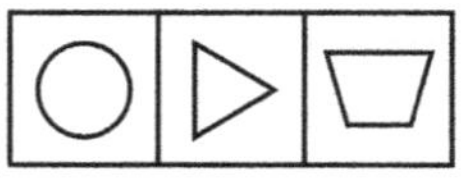 is the front view True or False

20. 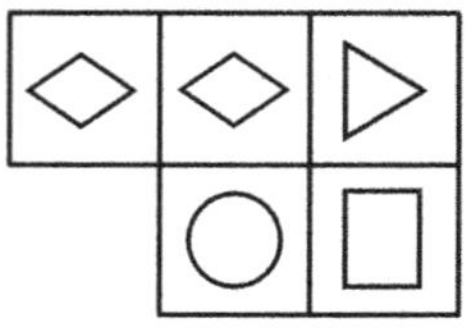 is the top view True or False

Matching

21. Match each side view with their given pile.

A. 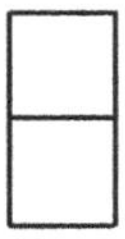

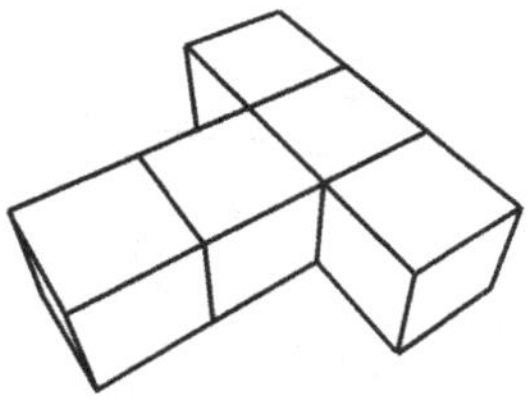

B. 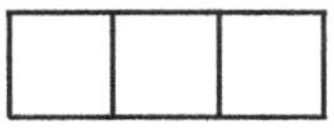

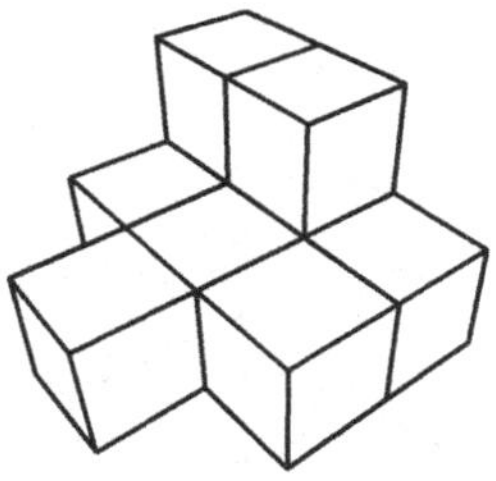

C. 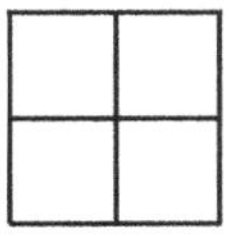

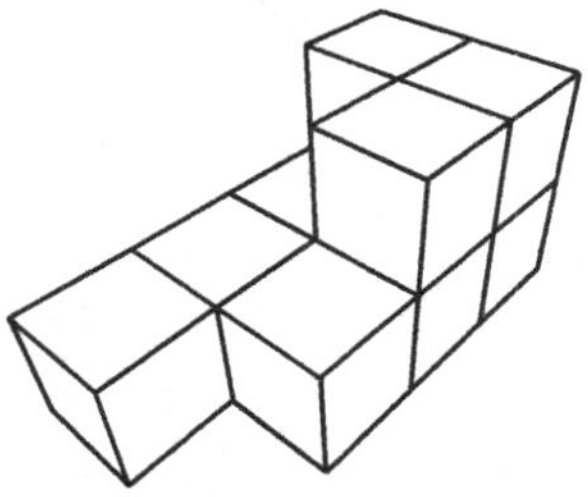

D. 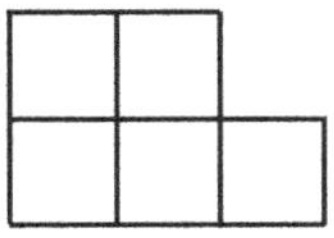

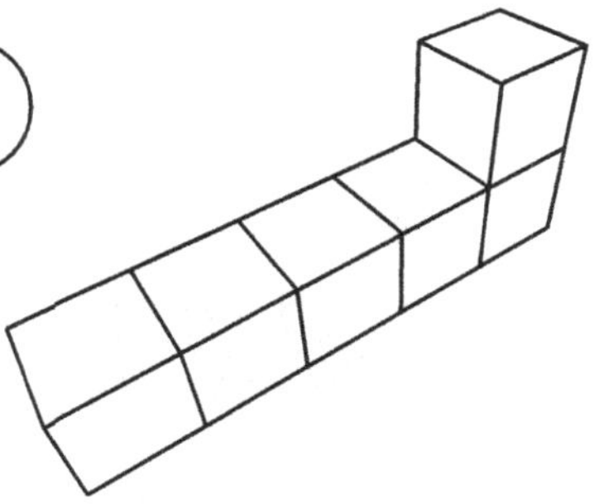

22. Cross out a single cube in each pile so that the structure perfectly matches the given top view, side view, and front view.

a.

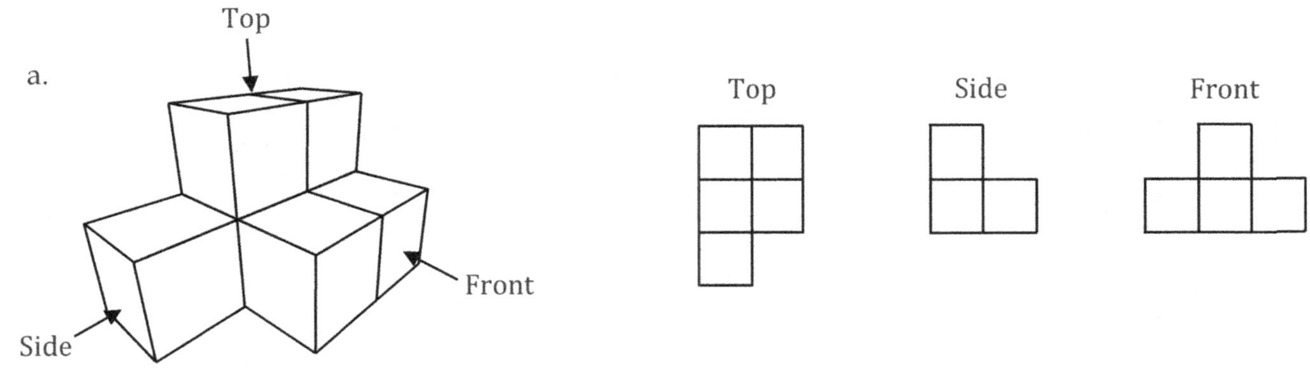

b.

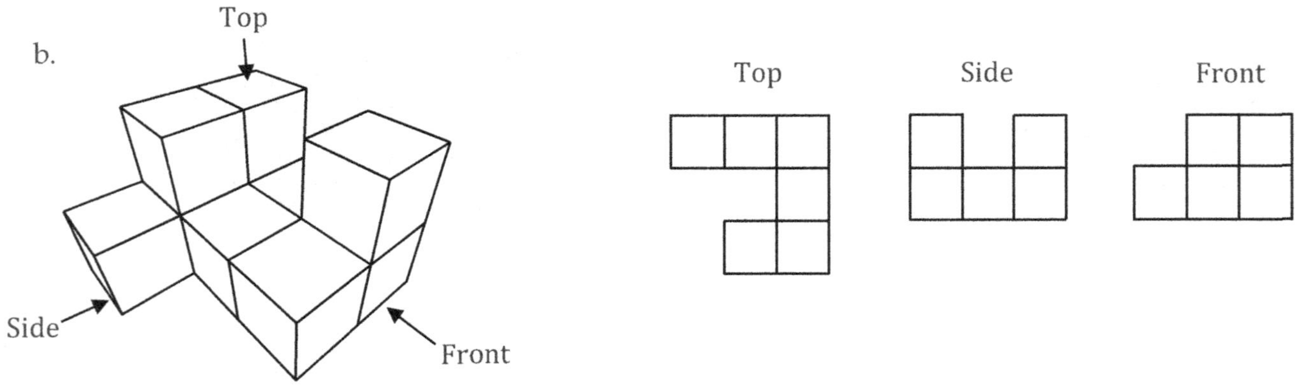

c.

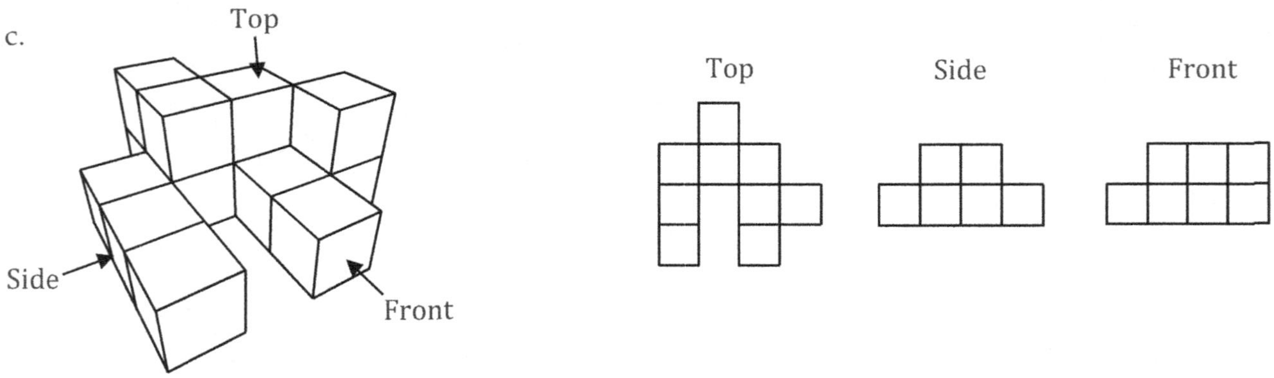

d.

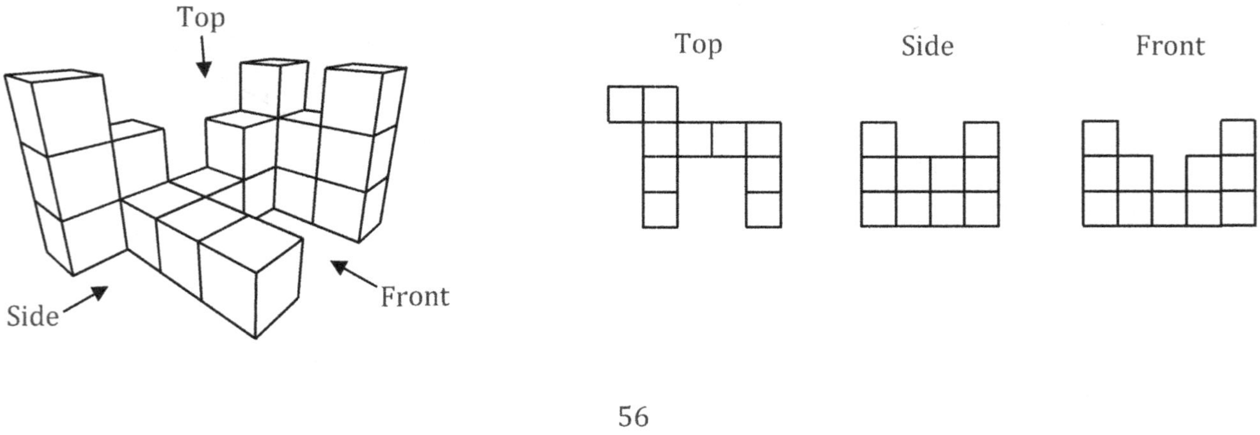

23. Shade in the squares that would indicate the correct front view, top view, and side view.

a.

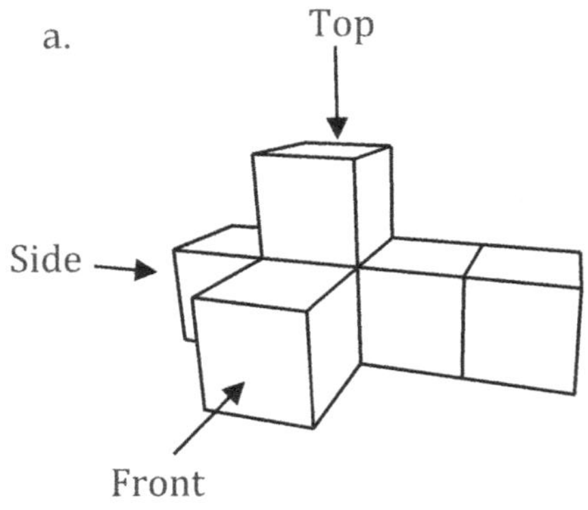

Front View | Top View | Side View

Total number of cubes in the pile: 6

b.

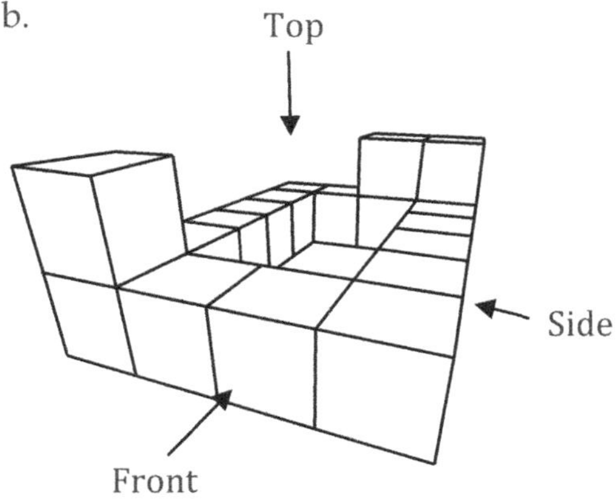

Front View | Top View | Side View

Total number of cubes in the pile: 19

c.

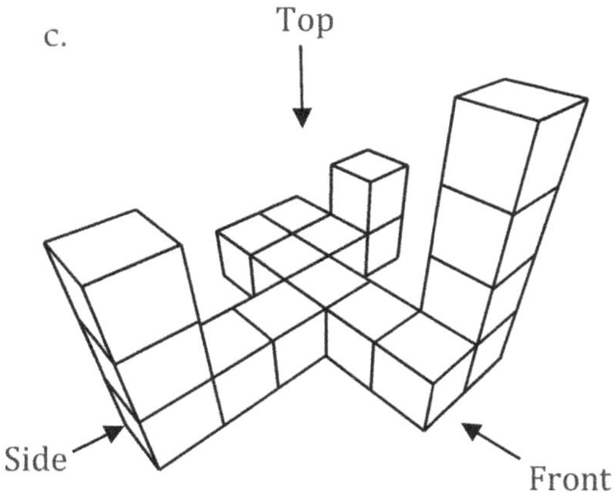

Front View | Top View | Side View

Total number of cubes in the pile: 18

24. Consider the pile of cubes shown below. This pile has exactly 7 cubes. However, from this given angle, only 6 cubes are visible. Exactly 1 cube is <u>completely</u> hidden when looking from this viewpoint.

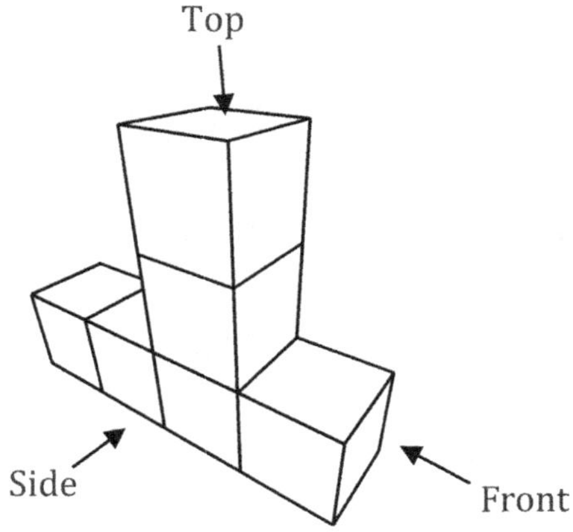

With this in mind, shade in the correct front view, top view, and side view.

Front View | Top View | Side View

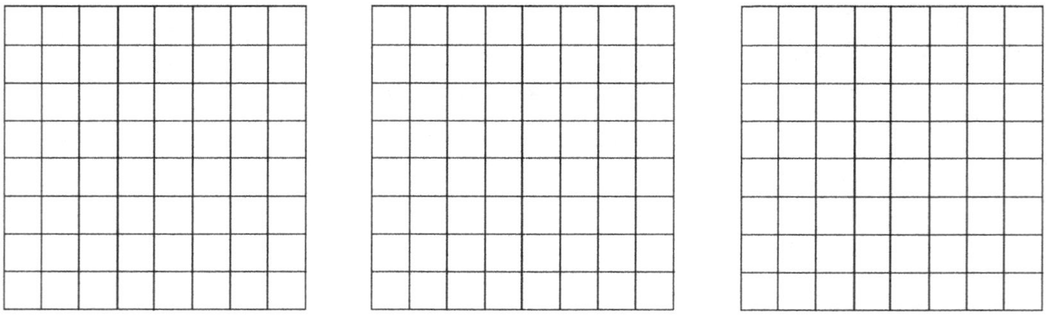

Total number of cubes in the pile: 7

End of Test

Next Page

Unit Test # 2

Estimated Duration: 1 hour 20 minutes

Unit Test

Spatial Reasoning with Blocks

50

Note, for every question, every cube is the same size and shape.
Some of cubes may be hidden from plain view and no cube is floating in the air.
If a cube is not laying on the ground, there must be another cube underneath holding it in place.

Multiple Choice

For questions 1 – 3, refer to the given structure

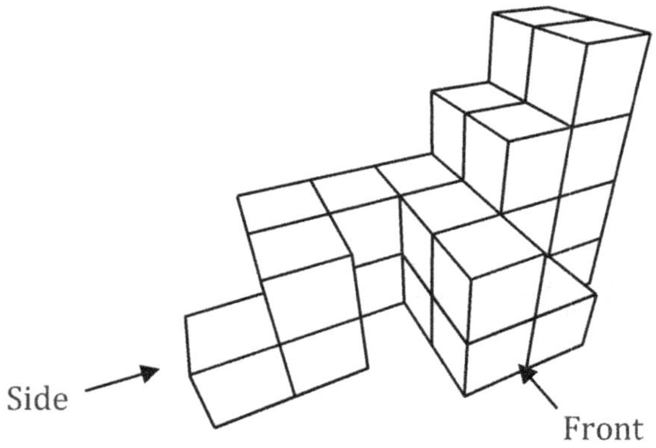

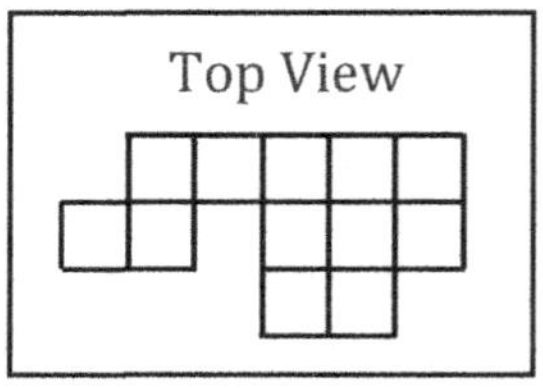

1. Circle the correct number of cubes in the pile.

a. 27 b. 28 c. 29 d. 30 e. 31 f. 32

2. Circle the correct side view for the pile.

a. 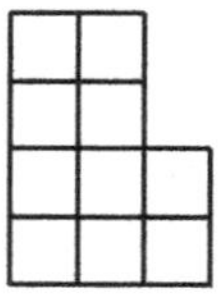 b. 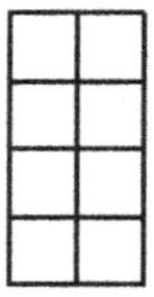 c. 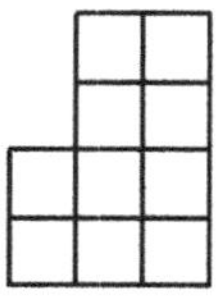 d. 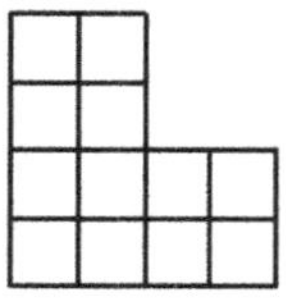 e. 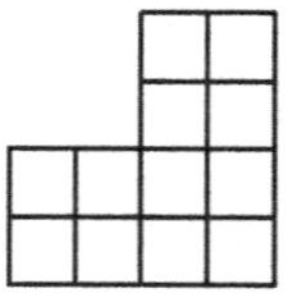

3. Circle the correct front view for the pile.

a. 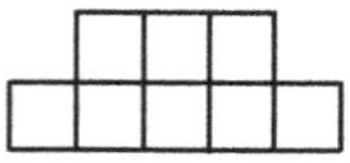 b. 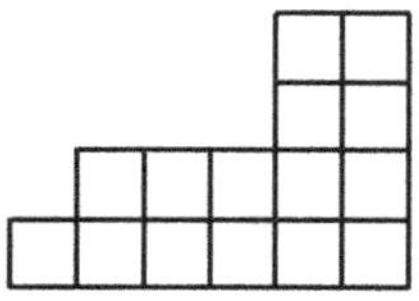 c. 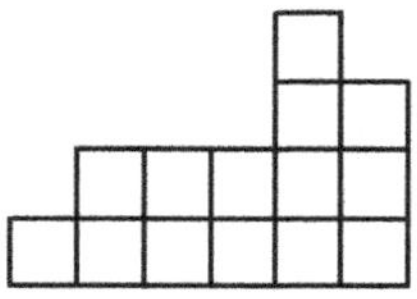 d. 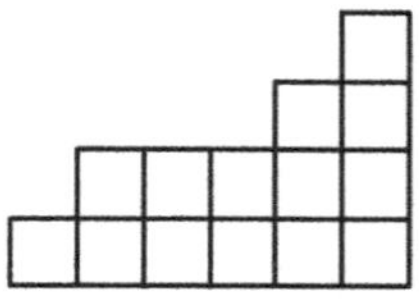

4. Choose the 3D-image that matches the specified views.

Top

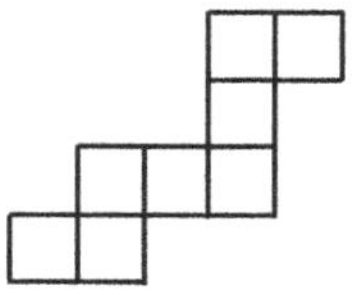

Front

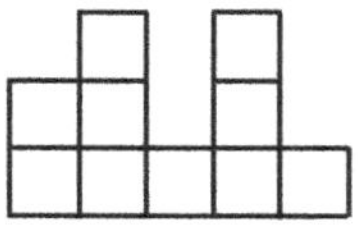

Side

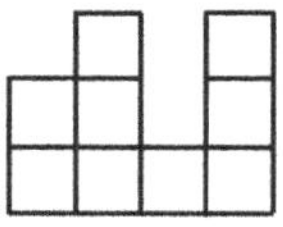

a.

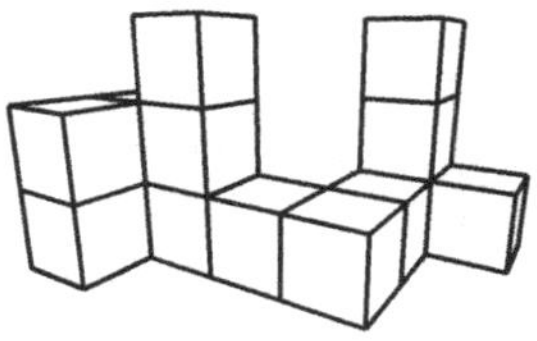

b.

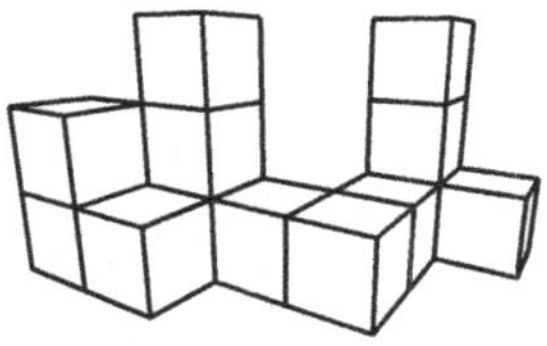

c.

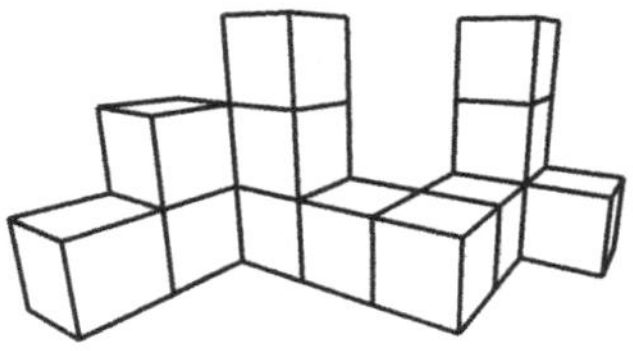

d.

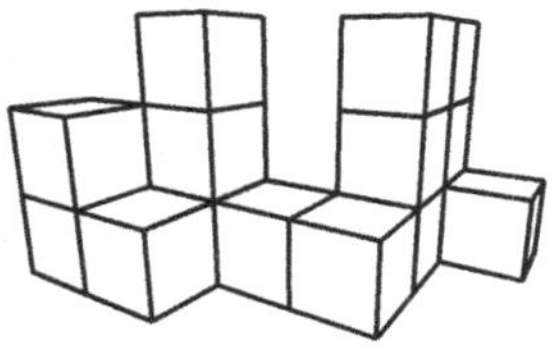

e.

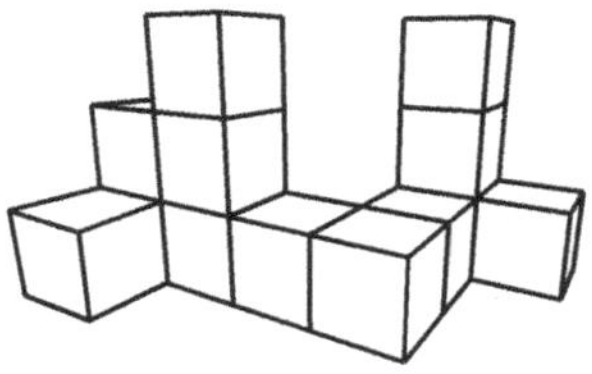

5. Choose the 3D-image that matches the specified views.

Top

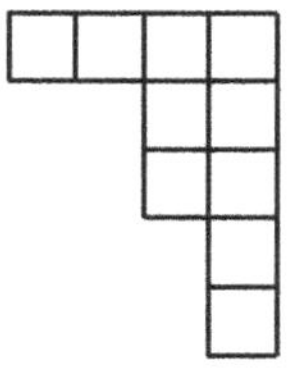

Front

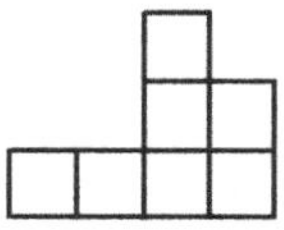

Side

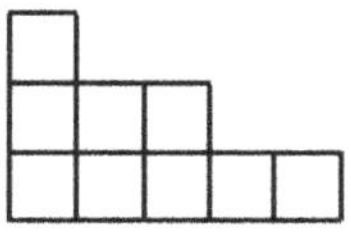

a.

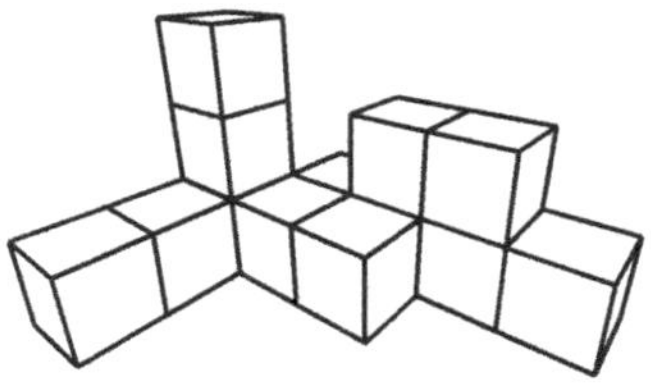

b.

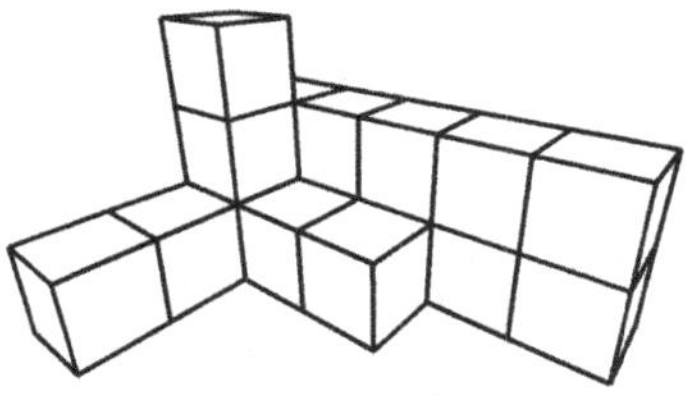

c.

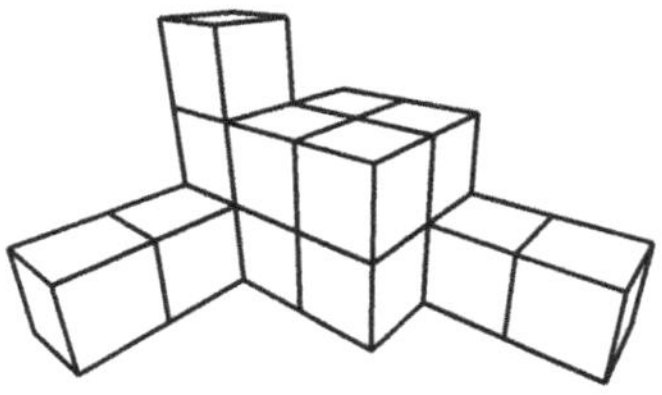

d.

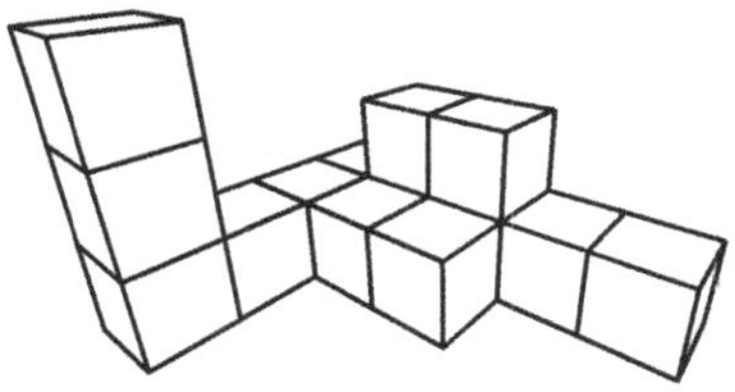

e.

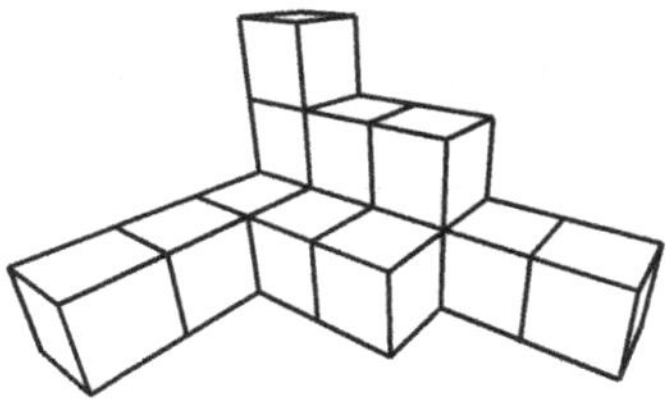

For questions 6 – 8, refer to the given structure

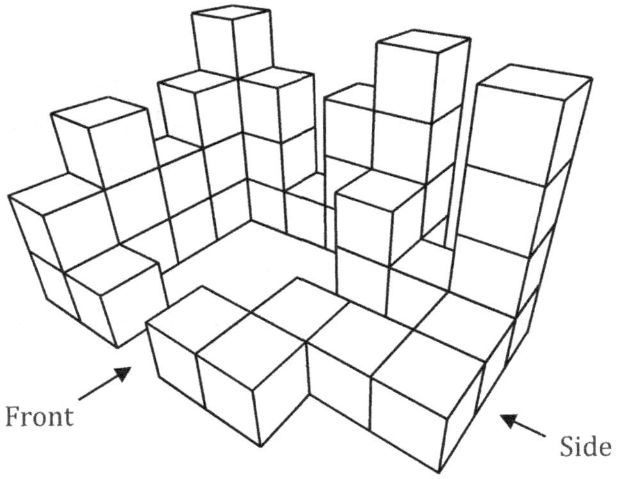

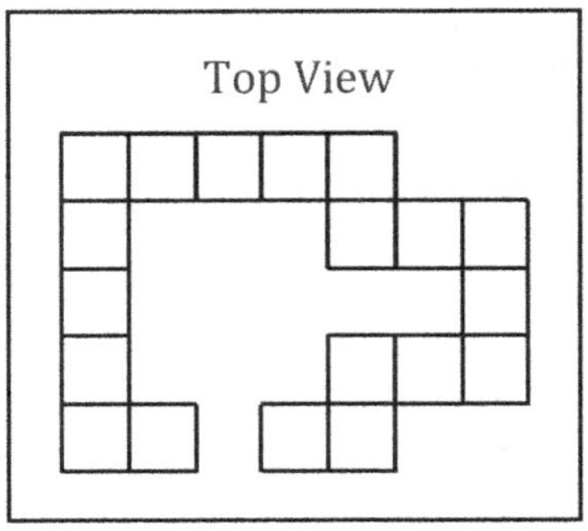

6. Circle the correct number of cubes in the pile.

a. 37 b. 38 c. 39 d. 40 e. 41 f. 42

7. Circle the correct side view for the pile.

a. 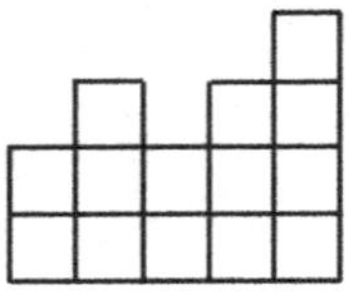

b. 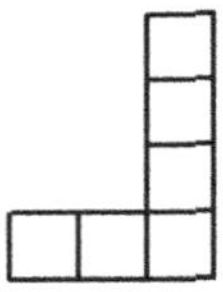

c. 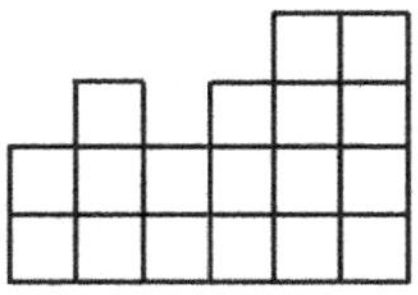

d. 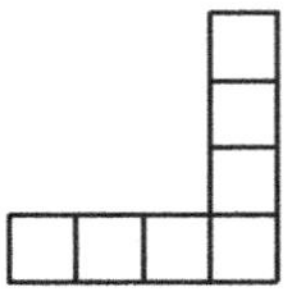

e. 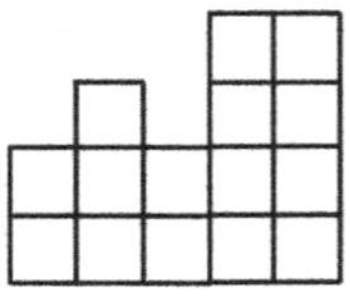

8. Circle the correct front view for the pile.

a. 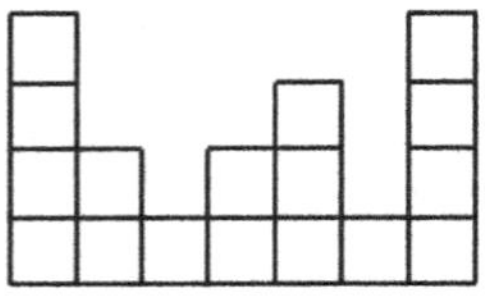

b. 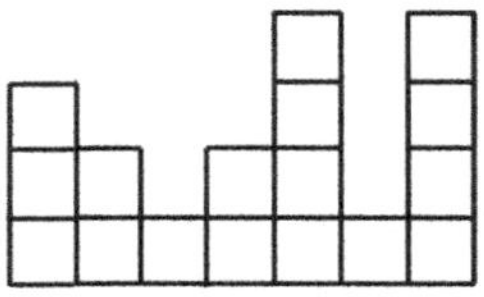

c. 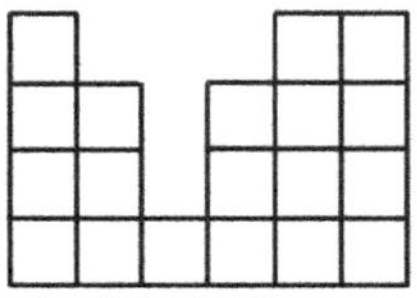

d. 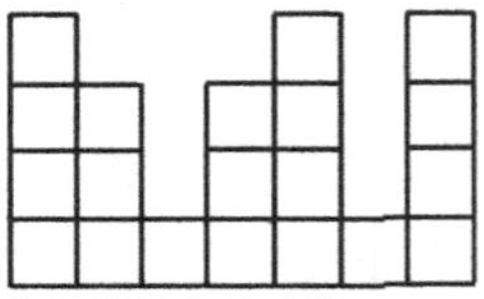

Recall two blocks are adjacent if the face of one block is touching the face of another block, even by a little bit. They are not adjacent if only their edges are touching. For questions 9 – 12, every block in each pile is the same size and shape.

9.

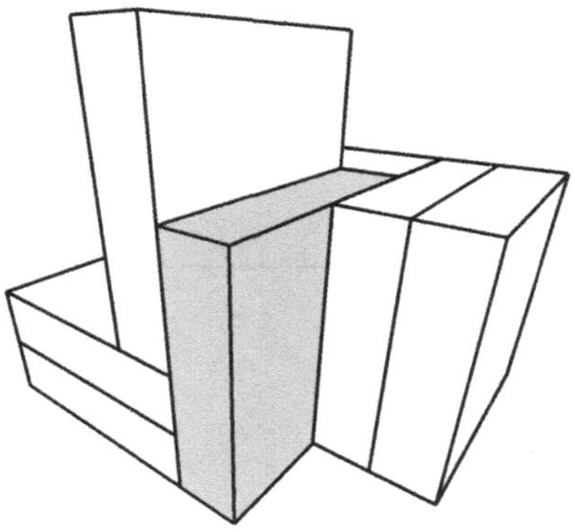

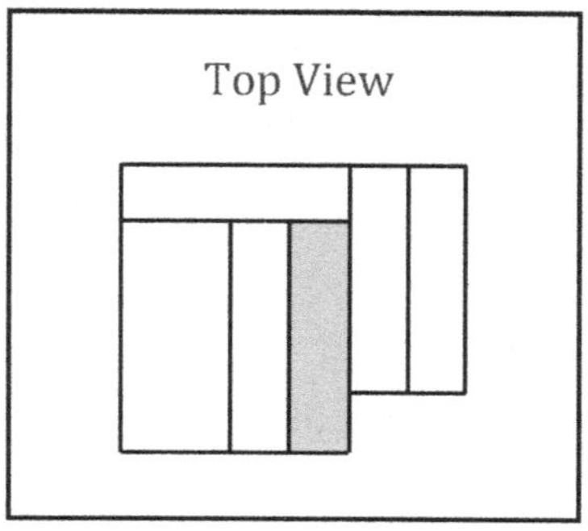

i. Circle the correct number of blocks in the pile.

a. 7 b. 6 c. 8 d. 5 e. 9 f. 4

ii. Circle the correct number of blocks adjacent to the shaded block in the pile.

a. 7 b. 6 c. 8 d. 5 e. 9 f. 4

10.

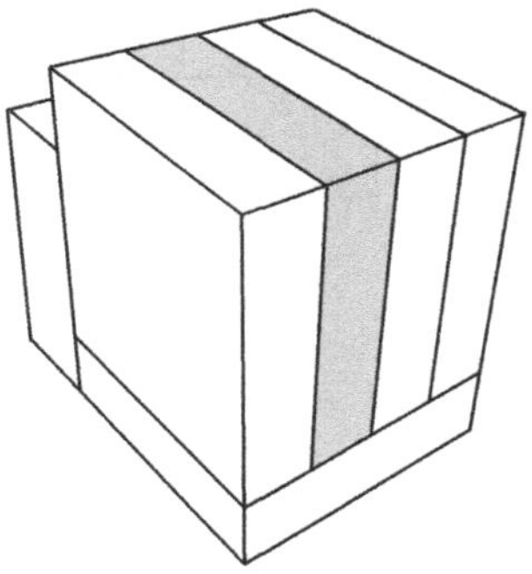

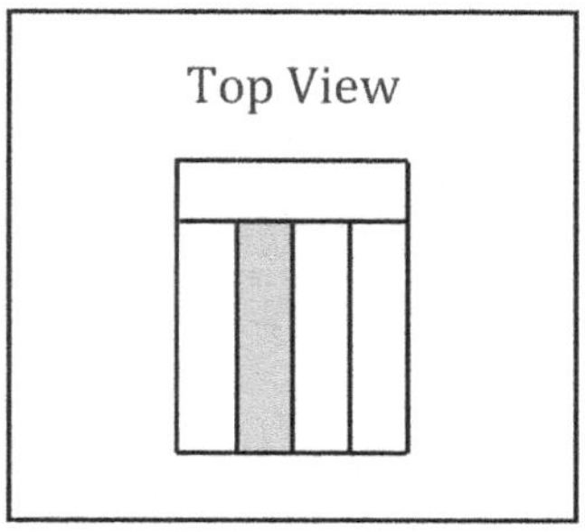

i. Circle the correct number of blocks in the pile.

a. 7 b. 6 c. 8 d. 5 e. 9 f. 4

ii. Circle the correct number of blocks adjacent to the shaded block in the pile.

a. 7 b. 6 c. 8 d. 5 e. 9 f. 4

11.

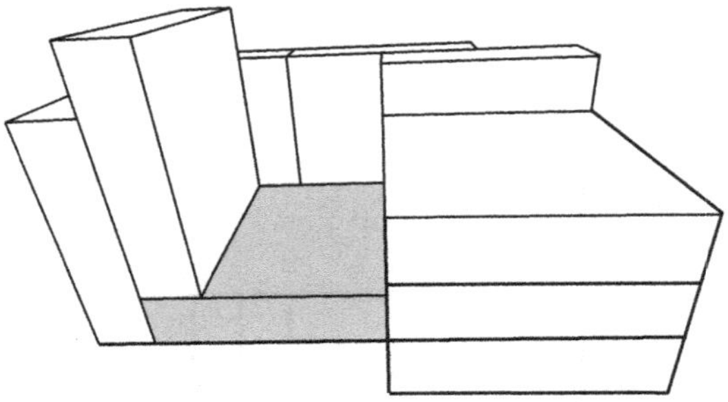

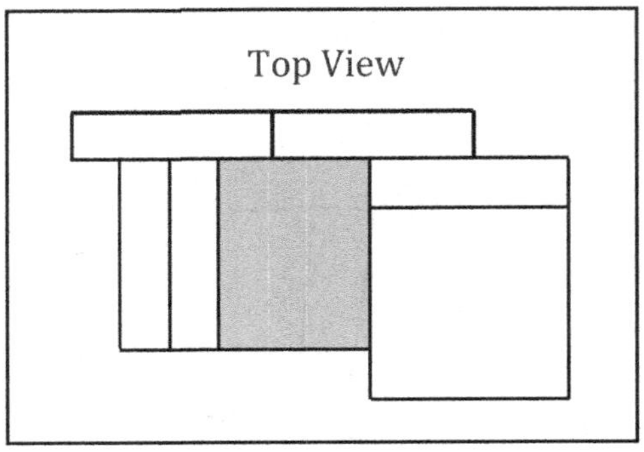

i. Circle the correct number of blocks in the pile.

a. 8 b. 7 c. 9 d. 6 e. 10 f. 5

ii. Circle the correct number of blocks adjacent to the shaded block in the pile.

a. 8 b. 7 c. 9 d. 6 e. 10 f. 5

12.

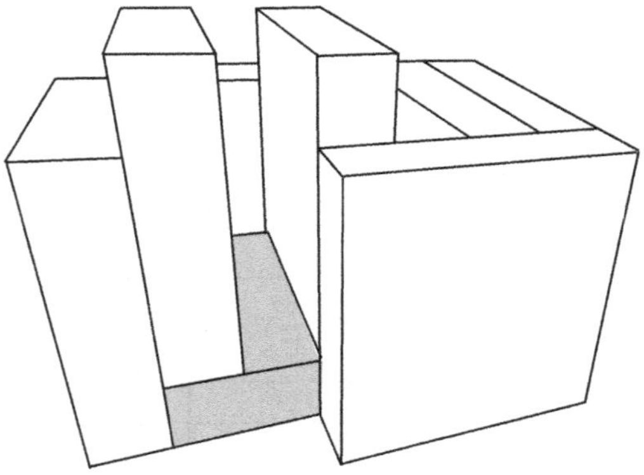

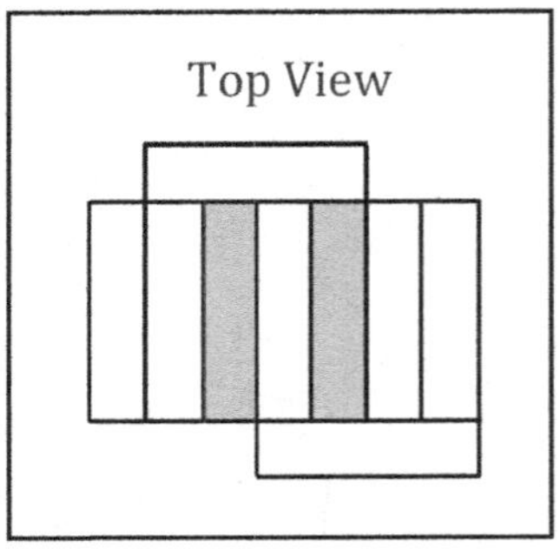

i. Circle the correct number of blocks adjacent to the shaded block in the pile.

a. 8 b. 7 c. 9 d. 6 e. 10 f. 5

ii. Circle the correct side view.

a. 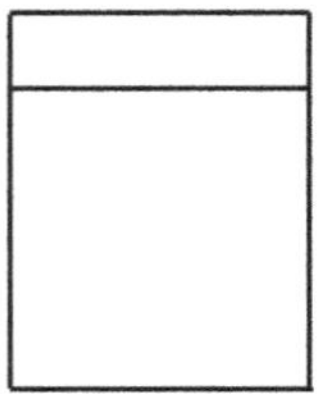 b. 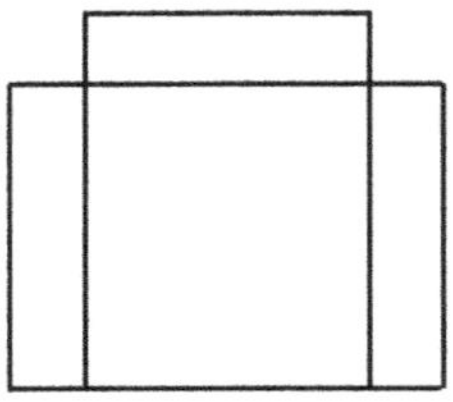 c. 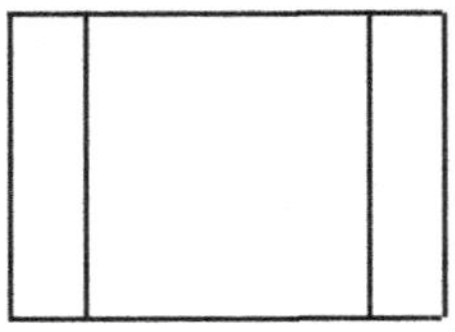 d. 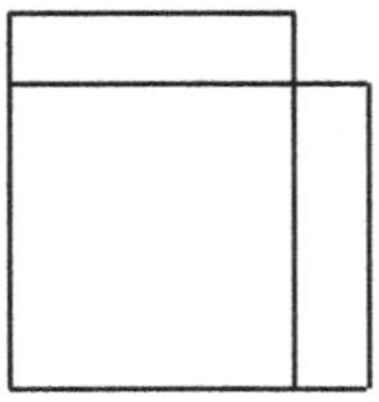

For questions 13 – 15, the bridge of cubes you come up with can be suspended in the air.

13. Circle the minimum number of cubes needed to bridge the two shaded cubes.

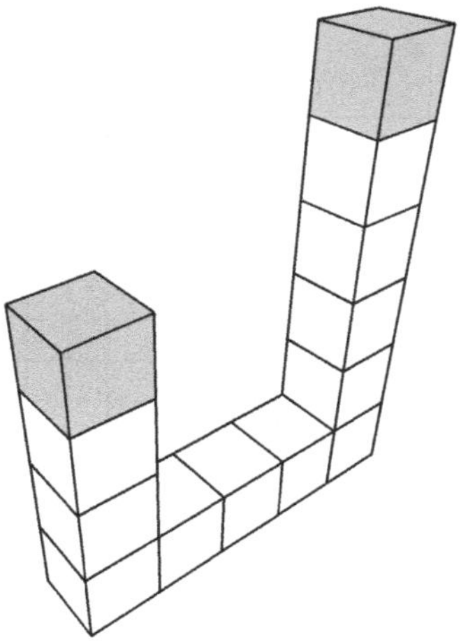

a. 7 b. 9 c. 11

d. 5 e. 3 f. 1

14. Circle the minimum number of cubes needed to bridge the two shaded cubes.

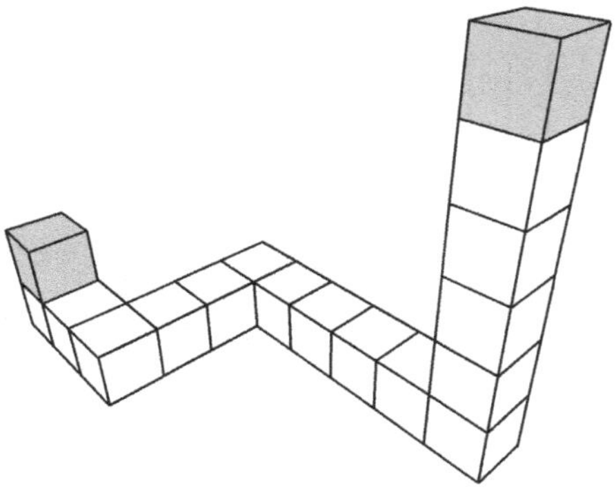

a. 11 b. 13 c. 15

d. 9 e. 7 f. 5

15. Circle the minimum number of cubes needed to bridge the two shaded cubes.

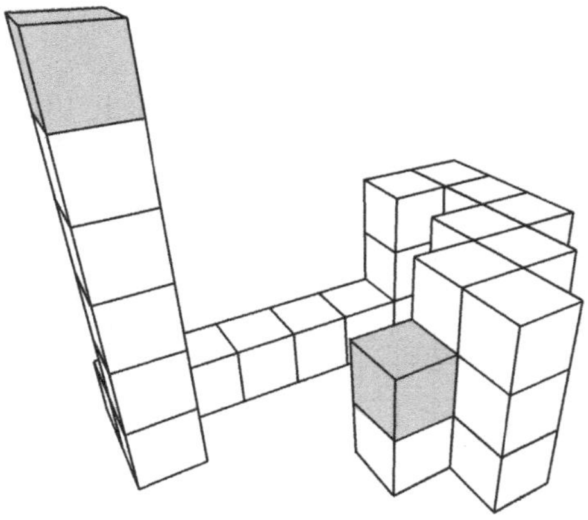

a. 10 b. 11 c. 12

d. 9 e. 8 f. 7

Shading

For questions 16 – 18, shade in the squares indicating the correct front view, top view, and side view.

16.

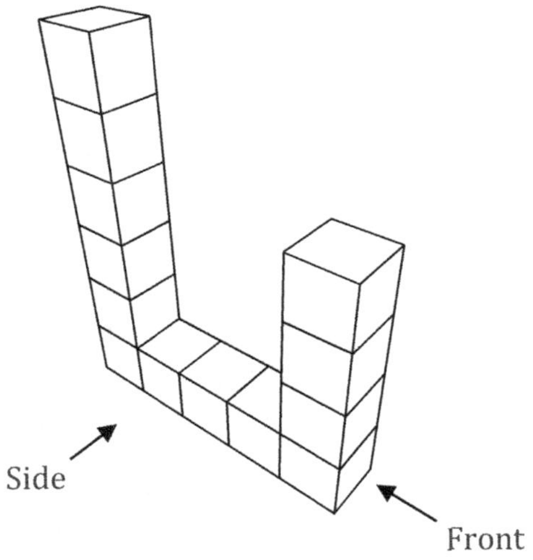

Front View

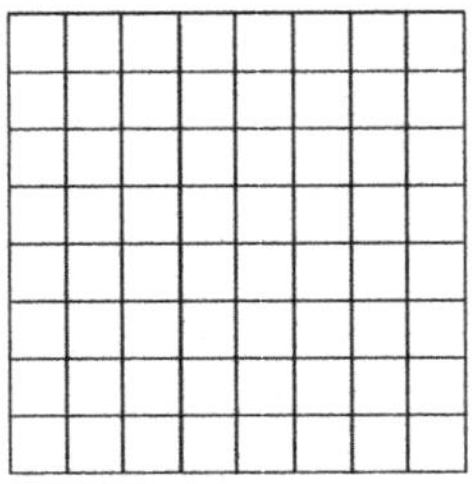

Top View

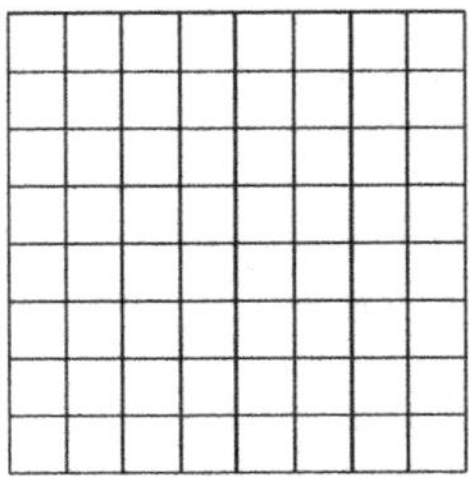

Side View

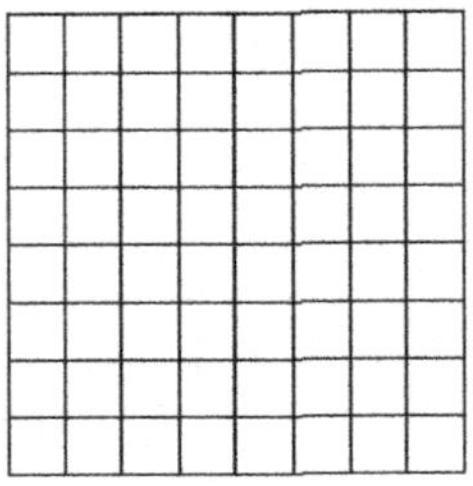

Total number of cubes in the pile: 13

/3

17.

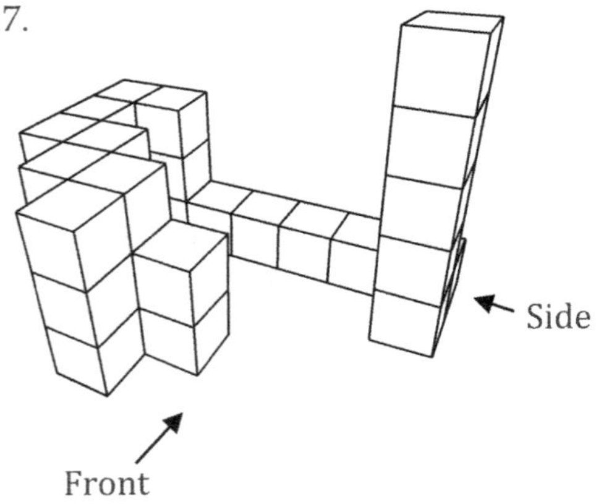

Front View

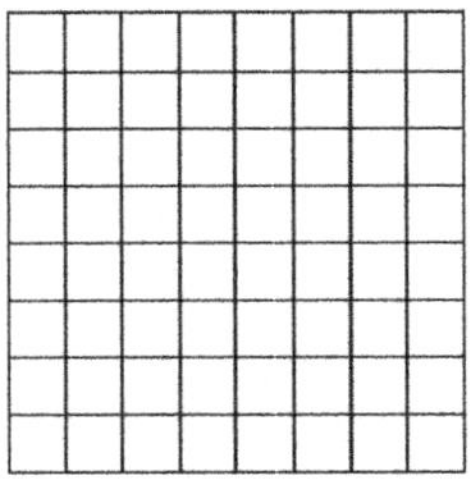

Top View

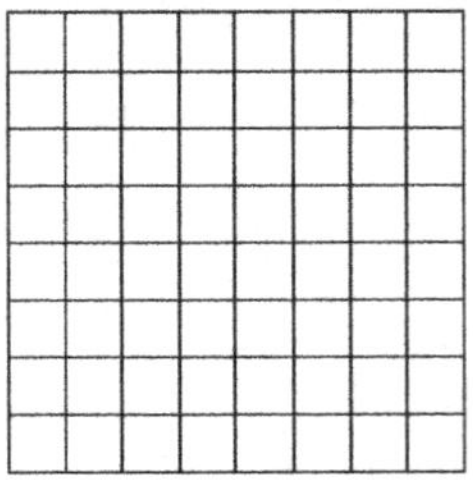

Side View

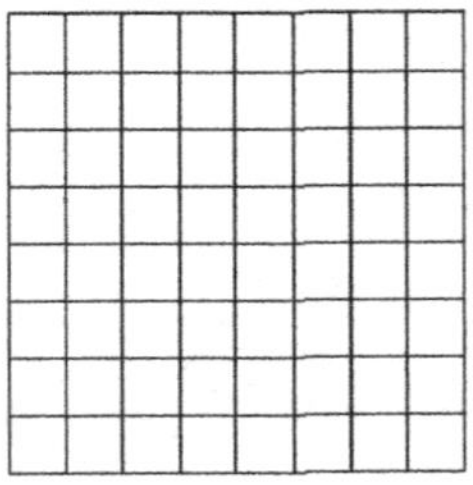

Total number of cubes in the pile: 36

/3

18.

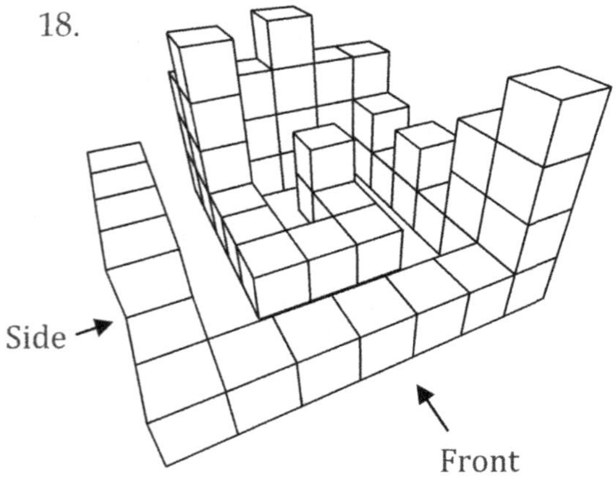

Front View

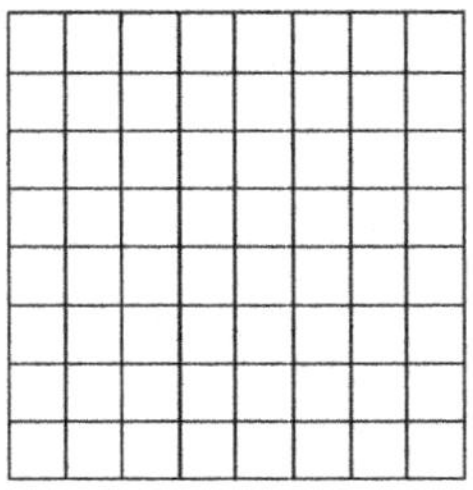

Top View

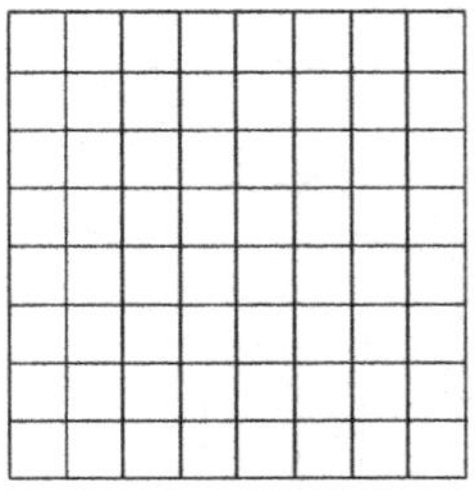

Side View

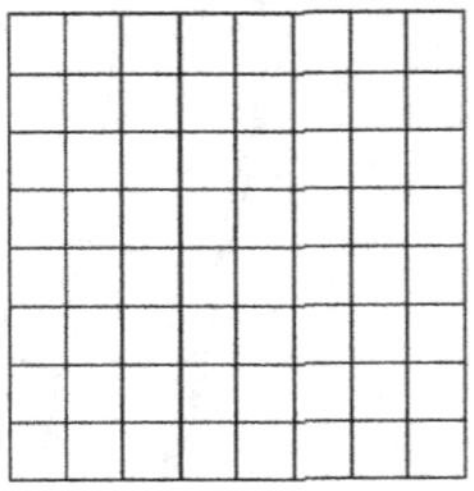

Total number of cubes in the pile: 56

/3

True or False

For questions 19 – 22, refer to the structure given below.

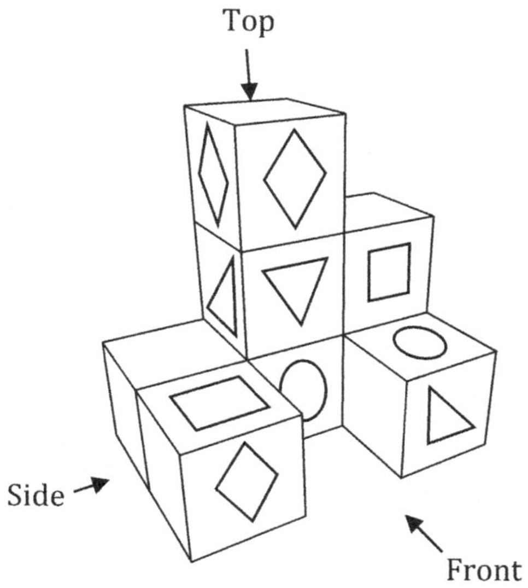

Circle the correct option.

19. 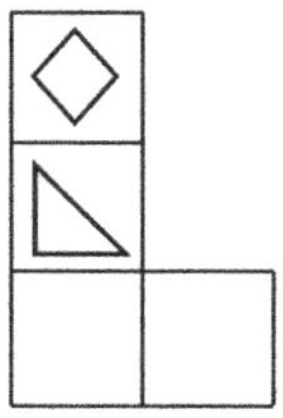 is the side view True or False

20. 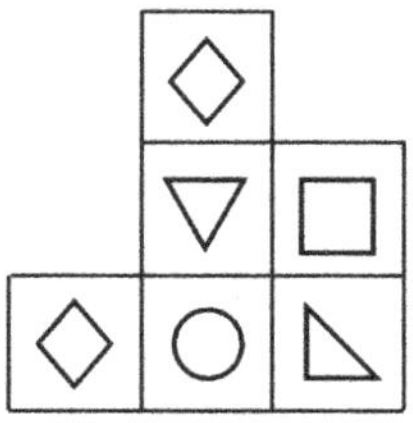 is the front view True or False

21. 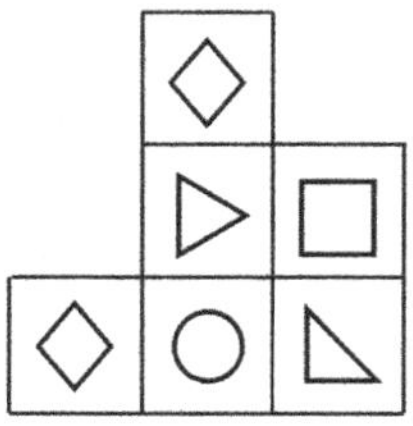 is the front view True or False

22. 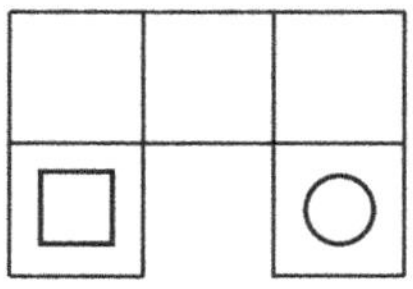 is the side view True or False

23. Cross out <u>two</u> cubes in the 3D pile so that the structure perfectly matches the given top view, side view, and front view.

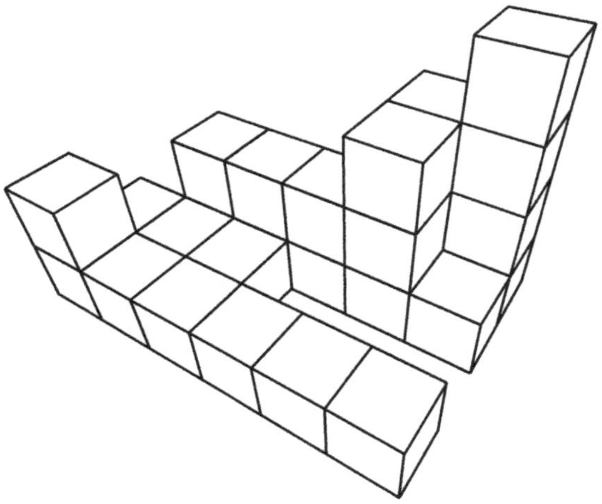

Top

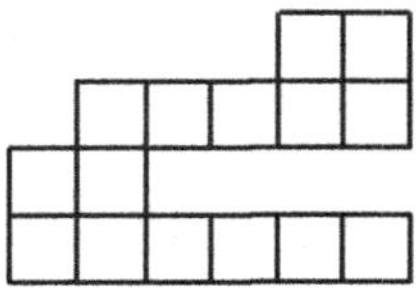

Side

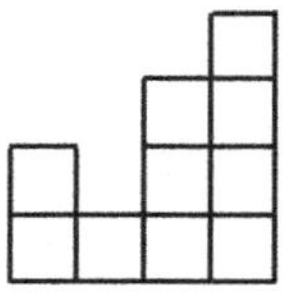

Front

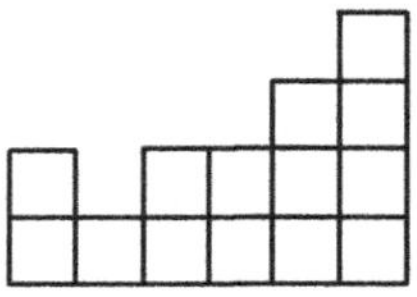

/1

24. Exactly one of the provided views is incorrect for the given solid. Cross out the wrong view.

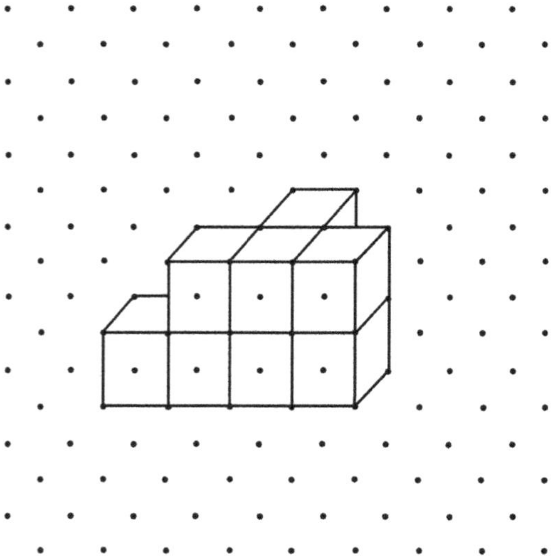

Top

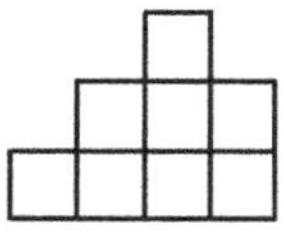

Side

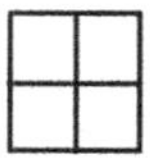

Front

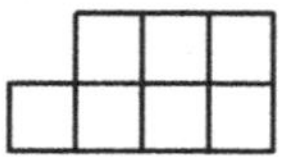

/1

25. Draw a 3D structure that matches the given views.

Side

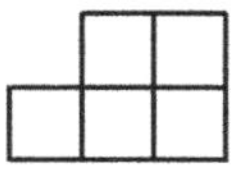

Front

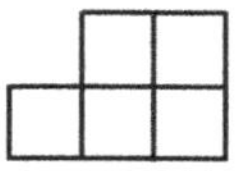

Top

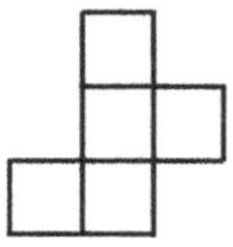

/3

26. Draw two <u>different</u> piles of cubes that <u>both</u> have the provided side view, front view, and top view.

Side

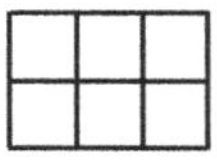

Front

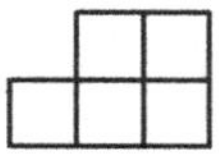

Top

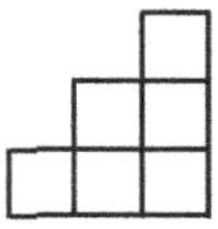

/4

27. A square pile is made using exactly 16 cubes. An extra cube is placed on top of each of the four corner cubes. Draw the correct front view, top view, and side view for this pile.

Front View

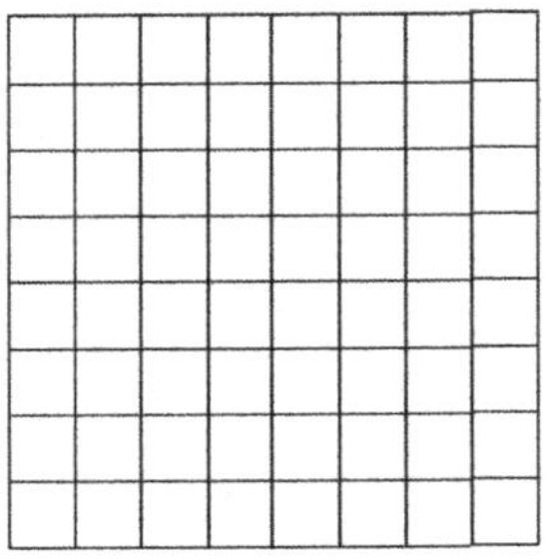

Top View

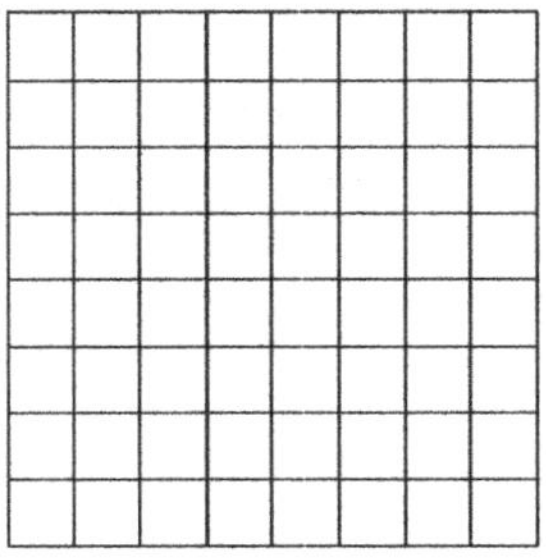

Side View

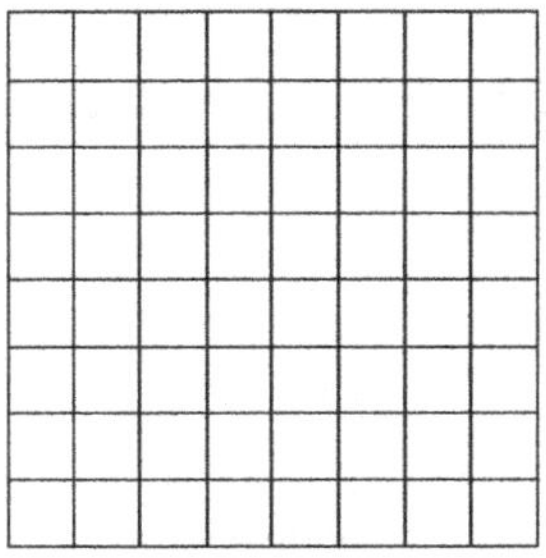

/3

28. Leo is looking straight at the front view of the structure given below. He uses his jetpack to float upward and forward until the structure is directly beneath him. Circle <u>exactly</u> what Leo will see.

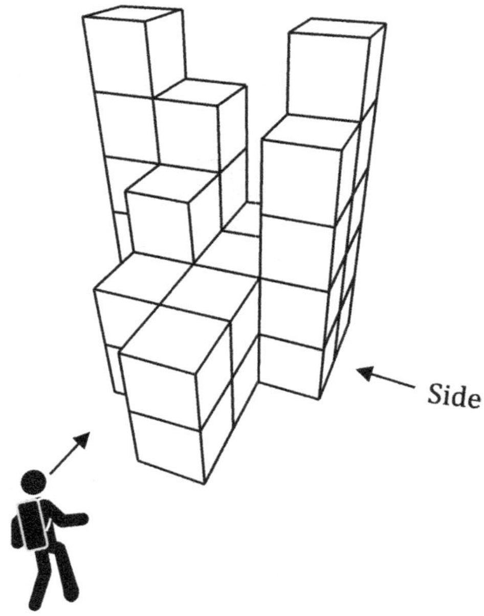

a. 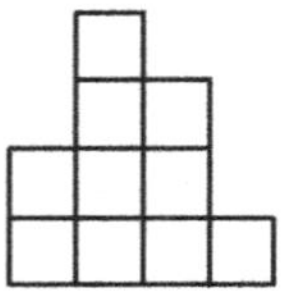

b. 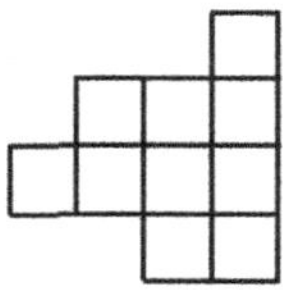

c. 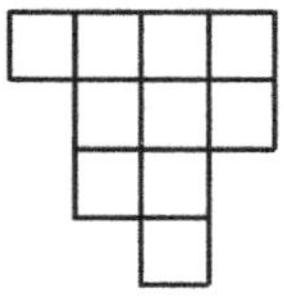

d. 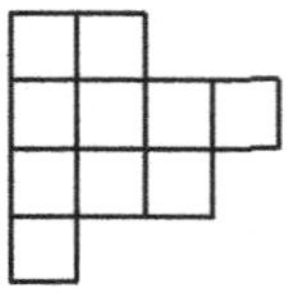

/1

29. A cube structure is placed on top of a building with a glass ceiling as shown below. There are two entrances to the building. One entrance is on the south side of the building the other entrance is on the east side of the building.

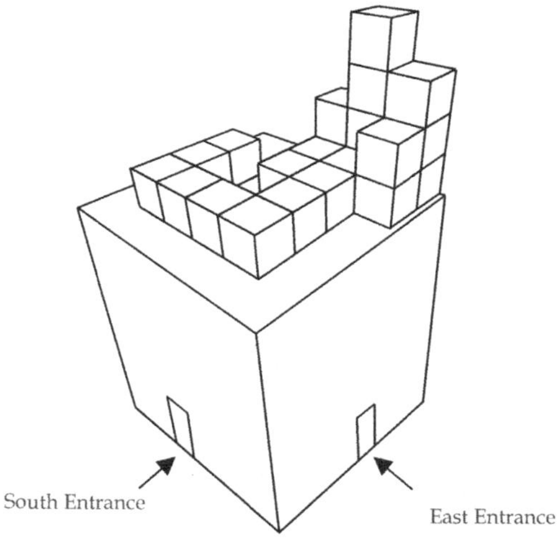

a. Bernard walks through the south entrance door and continues walking straight until he is directly underneath the cube structure. At that point, he looks upward. Circle exactly what he sees.

i.

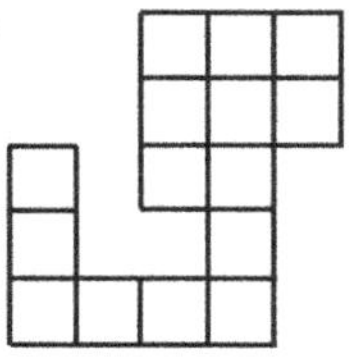

ii.

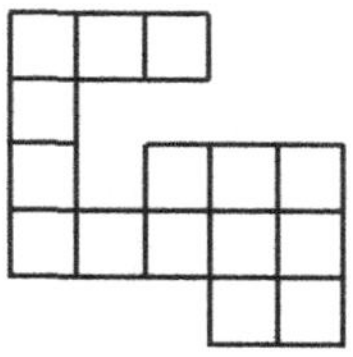

iii.

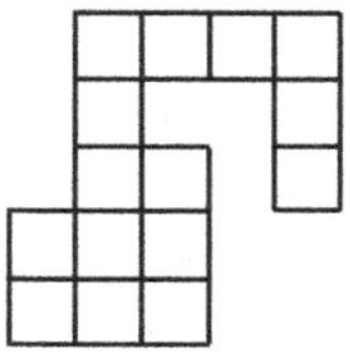

iv.

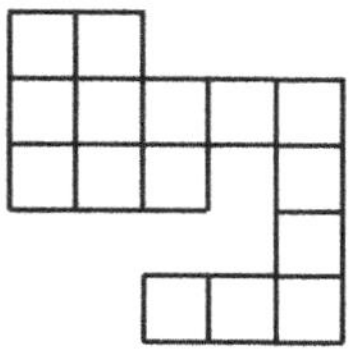

v.

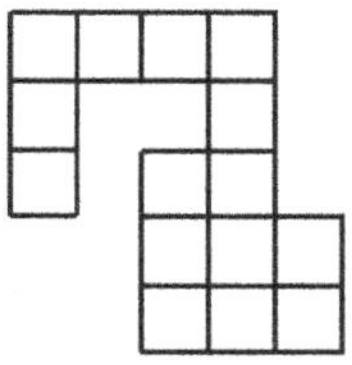

vi.

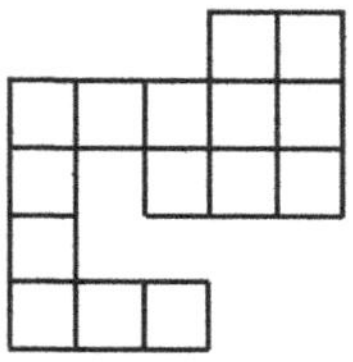

vii.

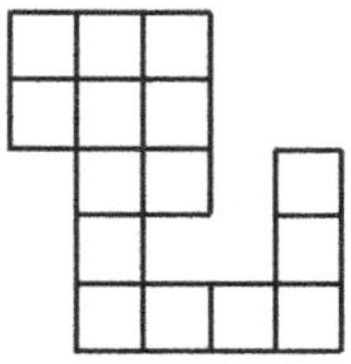

viii.

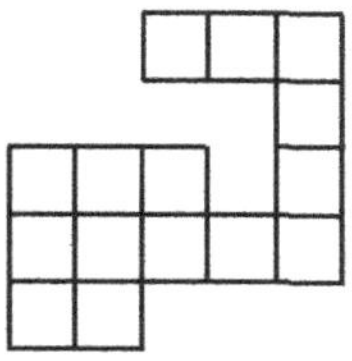

/1

b. Assume that Bernard walks through the east entrance instead. He walks straight until he is directly underneath the cube structure and then looks up. Shade in the exact view he sees.

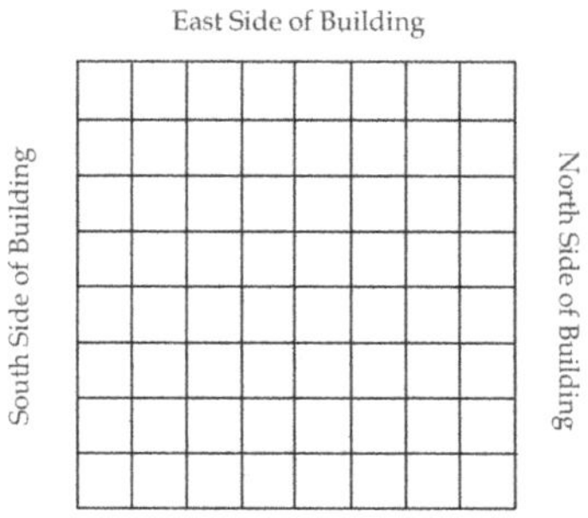

/2

30. A pile structure was broken up into the two piles shown below. The shaded sides used to be connected. Shade in the correct front view and top view for the original pile.

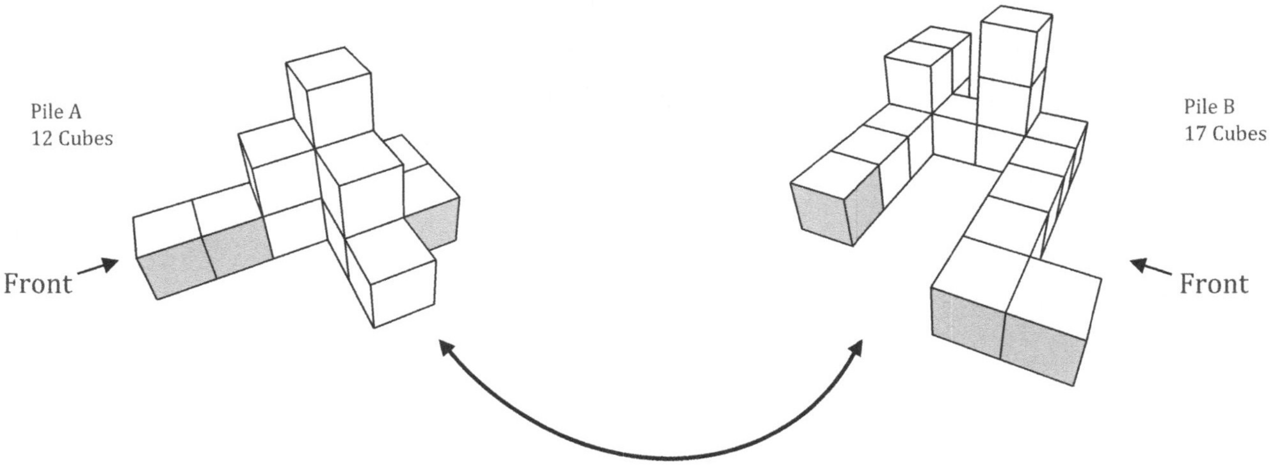

Front View

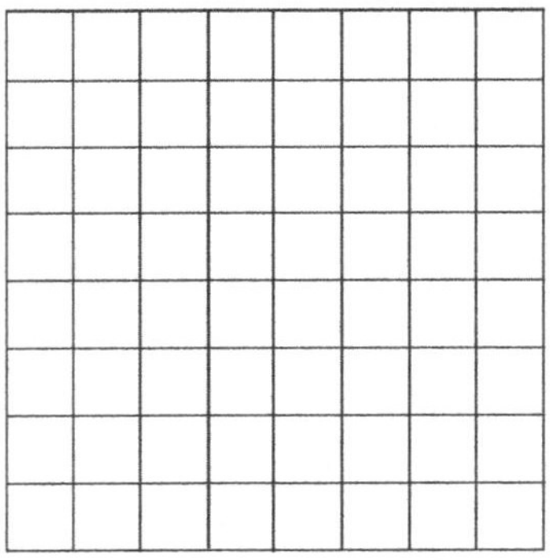

Top View

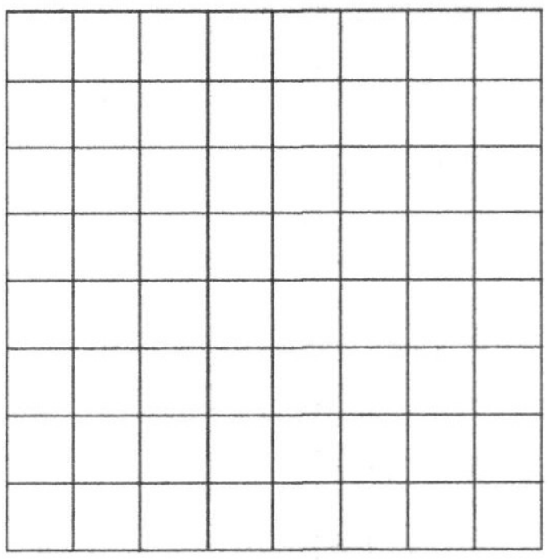

/2

End of Test

Answers

Adjacent Blocks

1.	3
2.	5
3.	4
4.	2
5.	5
6.	8
7.	6
8.	6
9.	7
10.	8
11.	5
12.	5
13.	7
14.	5
15.	5
16.	7
17.	5
18.	6
19.	6
20.	10
21.	7
22.	9

Counting Cubes

1.	7
2.	17
3.	120
4.	22
5.	15
6.	28
7.	14
8.	15
9.	21
10.	34
11.	30
12.	36
13.	34
14.	41
15.	64
16.	46
17.	90
18.	100
19.	97
20.	109
21.	82
22.	128
23.	134
24.	177

Connecting Cubes

1.	5
2.	6
3.	5
4.	7
5.	6
6.	17
7.	15
8.	12
9.	11
10.	18

Top View, Front View, and Side View Part 1

1. B
2. D
3. C
4. D
5. A
6. C
7. C
8. D
9. A
10. B

Top View, Front View, and Side View Part 2

Note: the exact placement of each shaded figure on the grid does not matter.
Furthermore, for the top view, the orientation of the shaded figure does not matter. All other rotations of the shaded figure provided for the top view are valid answers.

1.

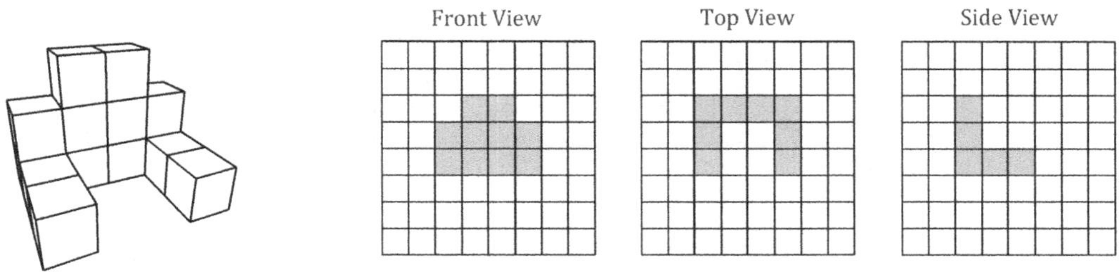

2.

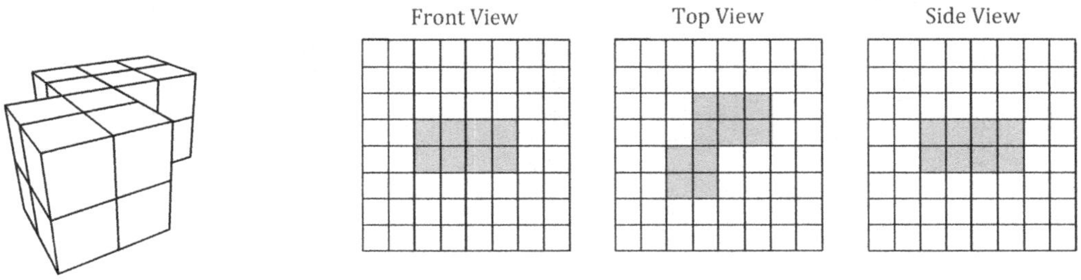

3.

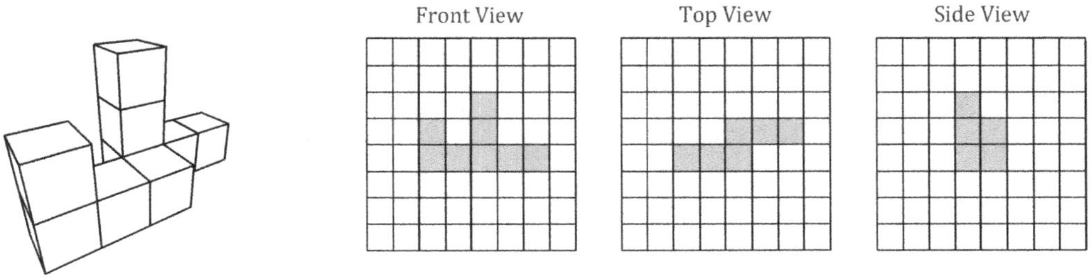

4.

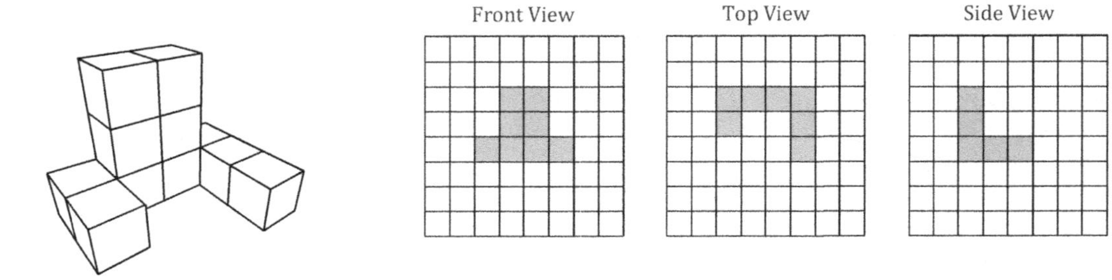
Front View
Top View
Side View

5.

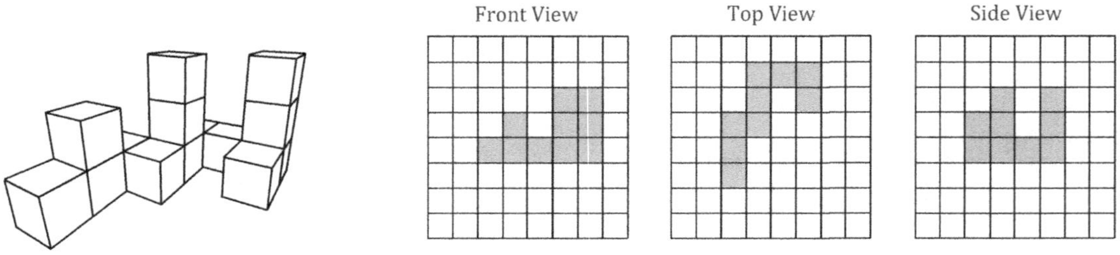
Front View
Top View
Side View

6.

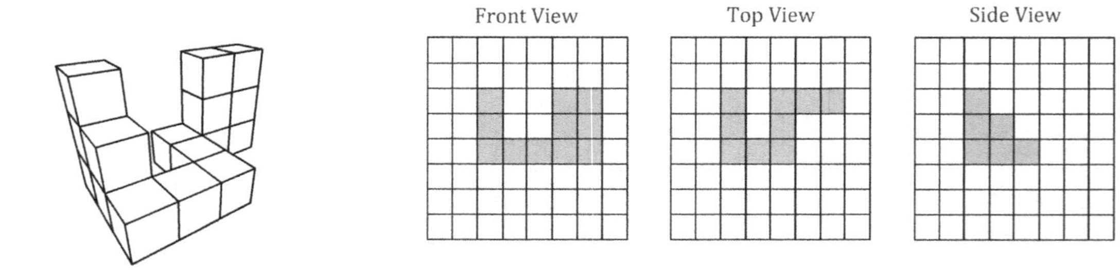
Front View
Top View
Side View

7.

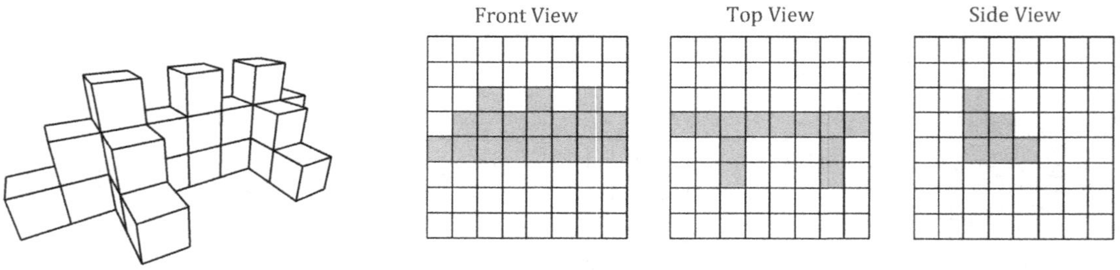
Front View
Top View
Side View

8.

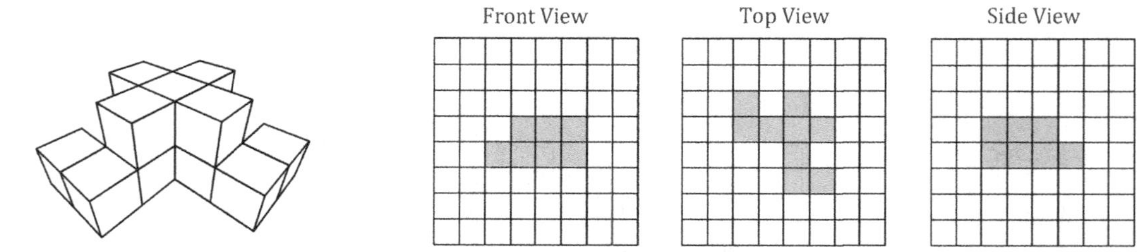
Front View
Top View
Side View

9.

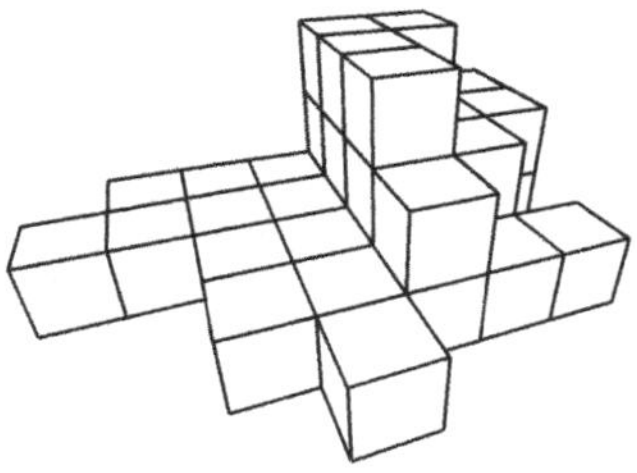

Front View

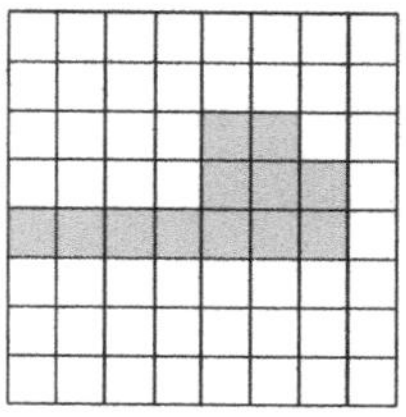

Top View

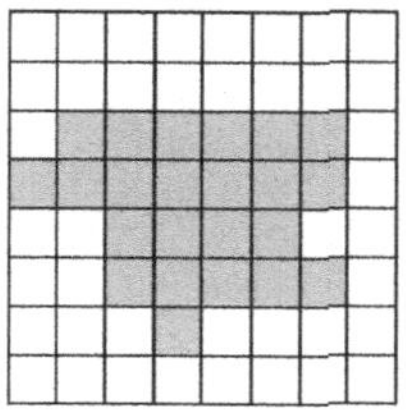

Side View

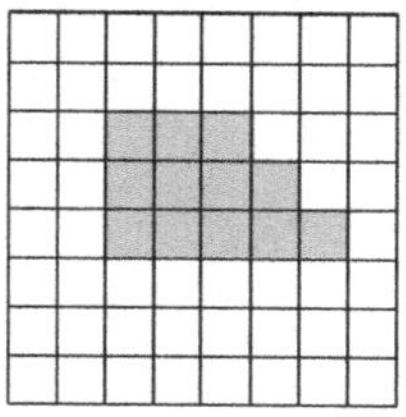

10.

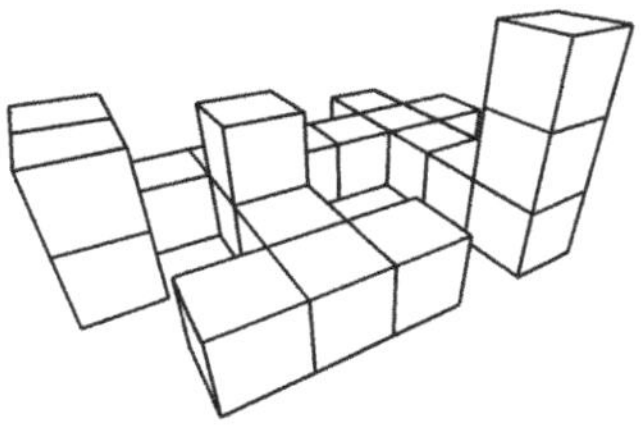

Front View

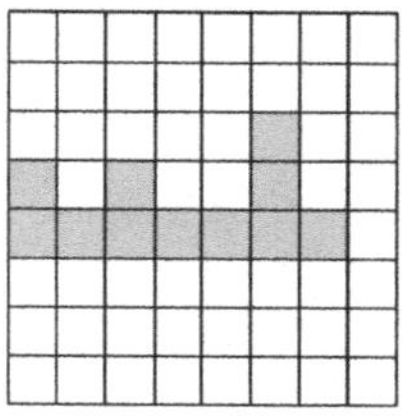

Top View

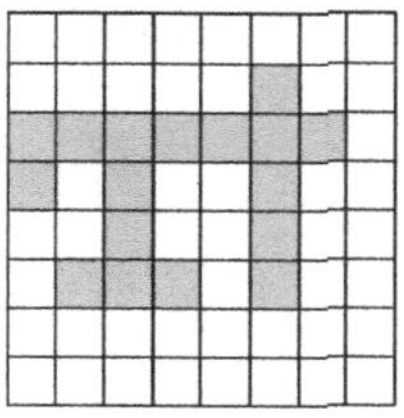

Side View

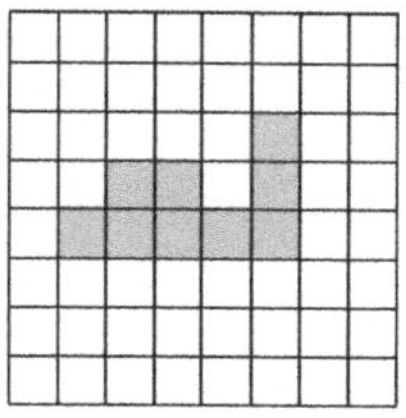

Top View, Front View, and Side View Part 3

1.	B	2.	E	3.	E
4.	D	5.	A	6.	A
7.	D	8.	C	9.	D
10.	A				

Unit Test # 1 Answers

1.	C	2.	B	3.	A
4.	B	5.	D	6.	D
7.	B	8.	A	9.	D
10.	A	11.	G	12.	A
13.	C	14.	A	15.	C
16.	True	17.	False	18.	False
19.	False	20.	True		

21.

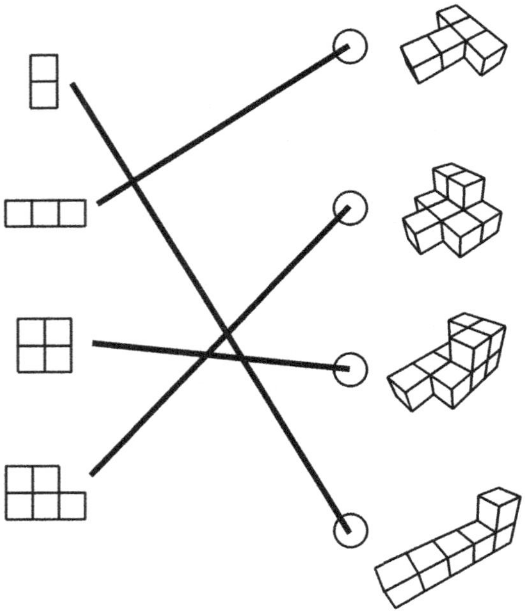

22.

a.

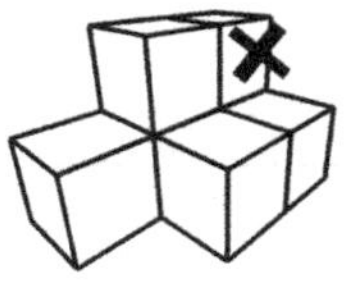

b.

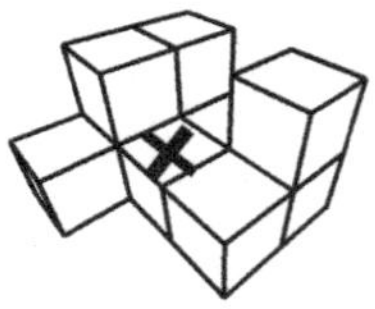

c.

d.

23.

a.

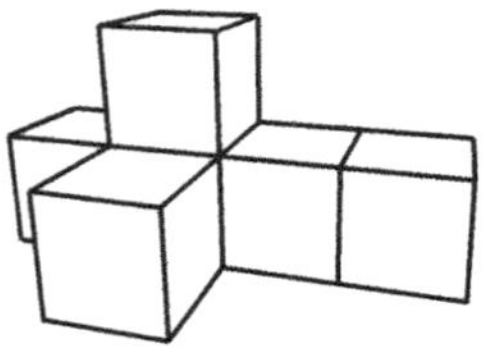

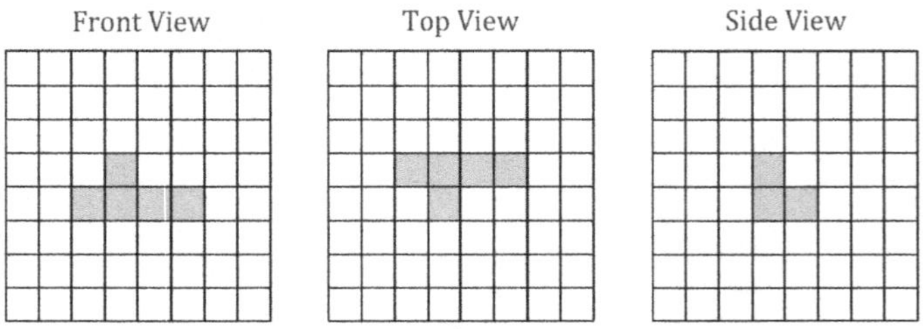

Or any other rotation

b.

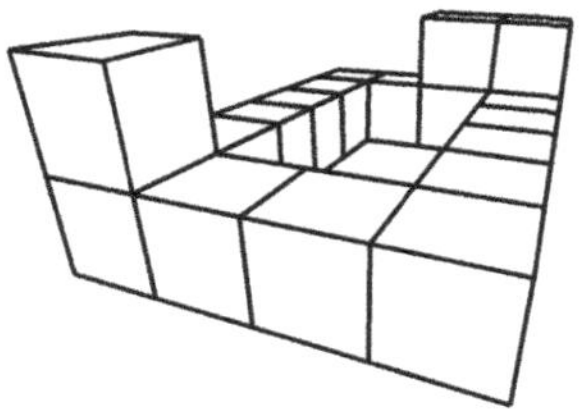

Front View

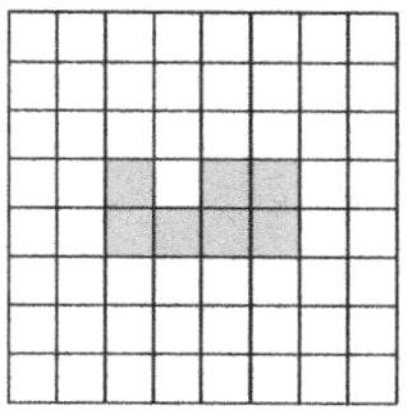

Top View

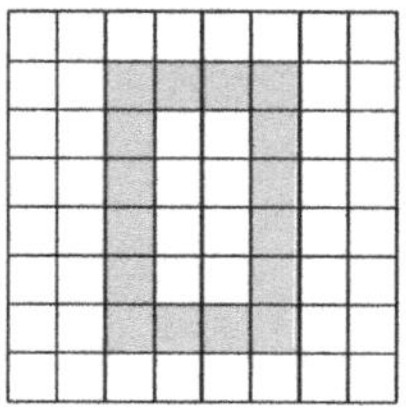

Or any other rotation

Side View

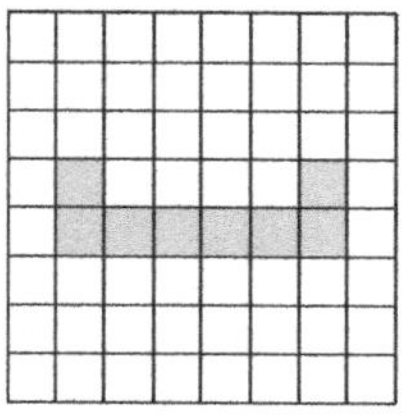

c.

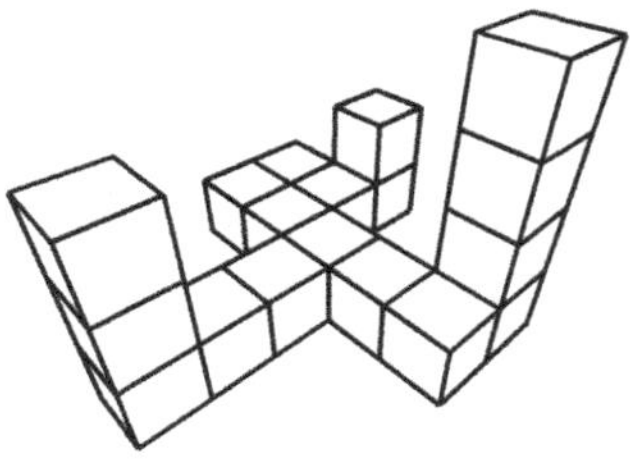

Front View

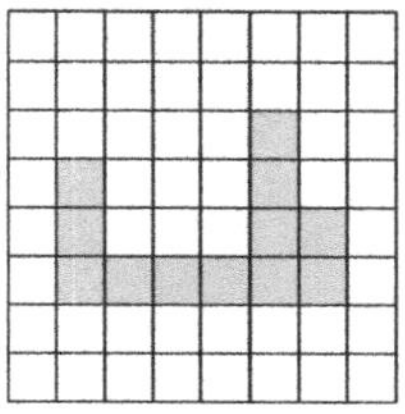

Top View

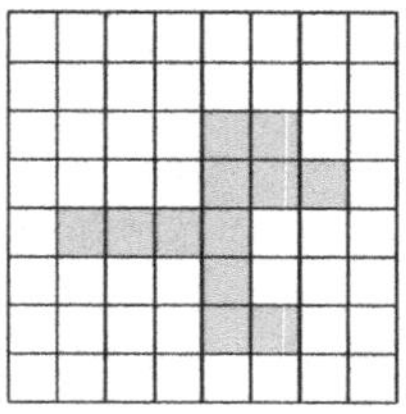

Or any other rotation

Side View

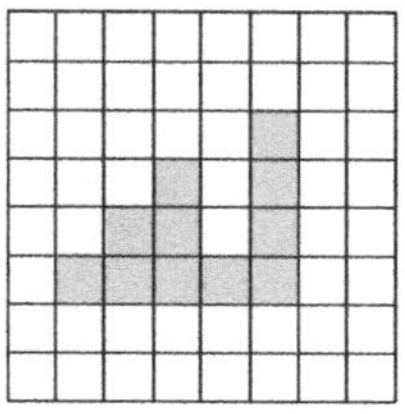

24.

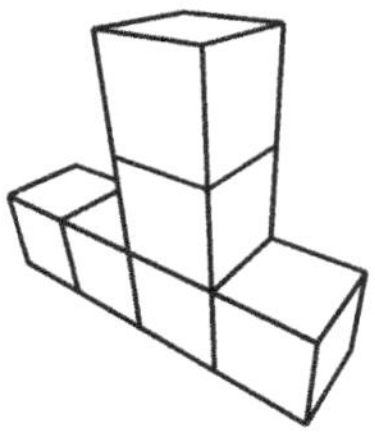

Front View

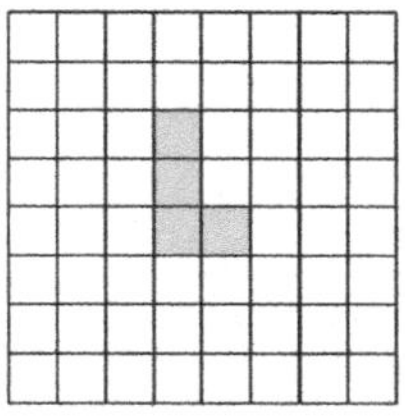

Top View

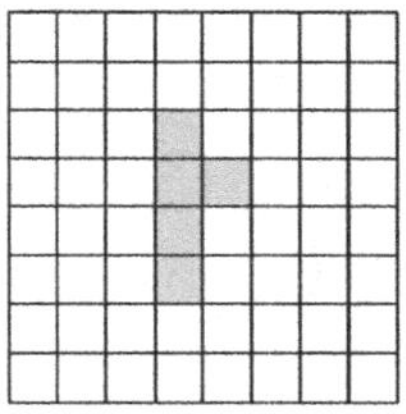

Or any other rotation

Side View

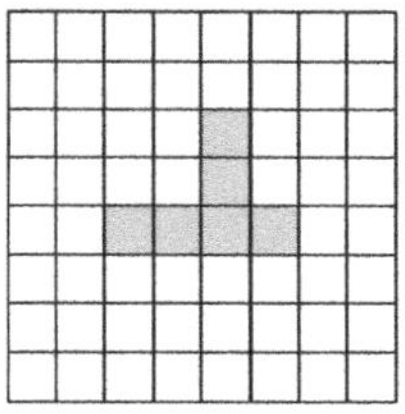

Unit Test # 2 Answers

1.	B	2.	A	3.	D
4.	B	5.	C	6.	C
7.	E	8.	D	9i.	A
9ii.	D	10i.	B	10ii.	F
11i.	C	11ii.	D	12i.	D
12ii.	B	13.	D	14.	B
15.	E				

16.

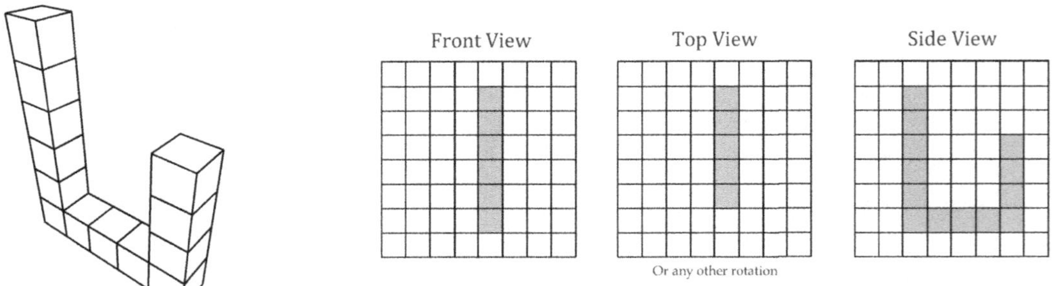

17.

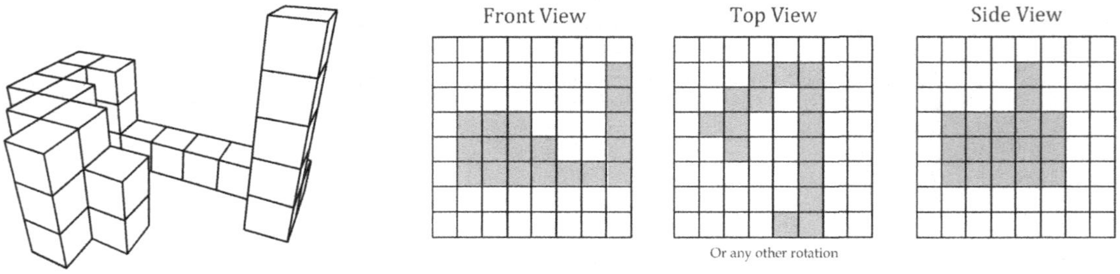

18.

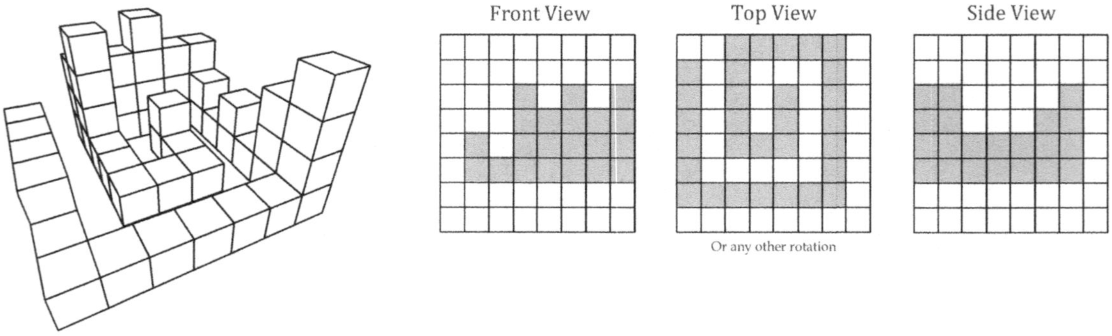

19. False

20. True

21. False

22. False

23.

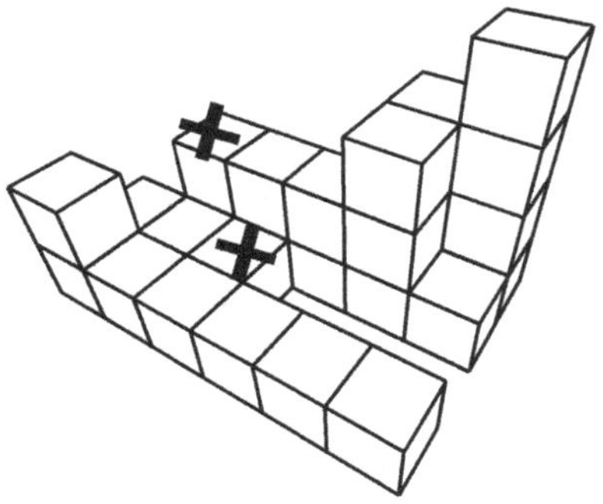

24.

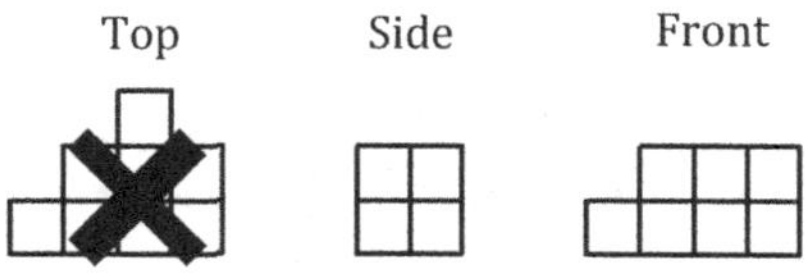

25. **Answers may vary.** Below are two examples.

Example #1

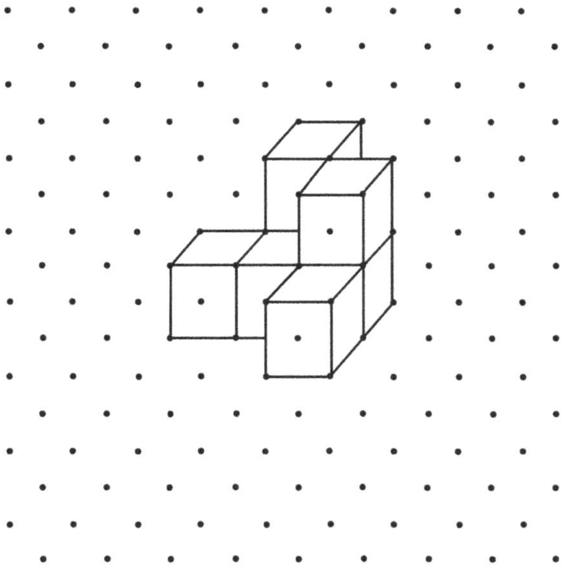

Example #2

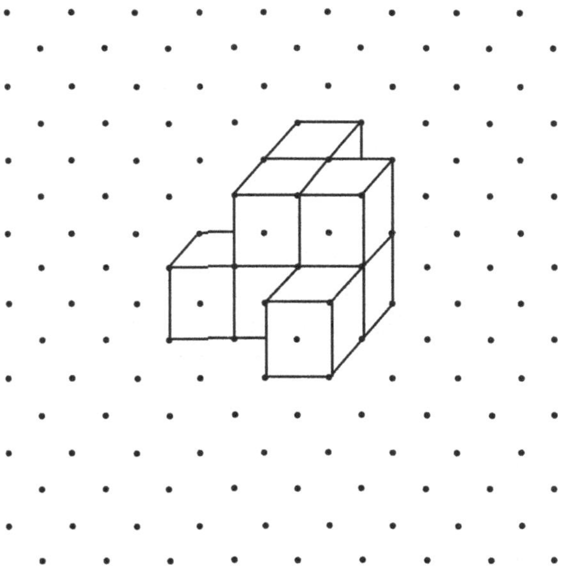

Marking Scheme

1 mark given if the provided Side View matches the structure drawn.
1 mark given if the provided Front View matches the structure drawn.
1 mark given if the provided Top View matches the structure. Total of 3 marks.

26. **Answers may vary.** An example of a correct answer is provided below.

Example

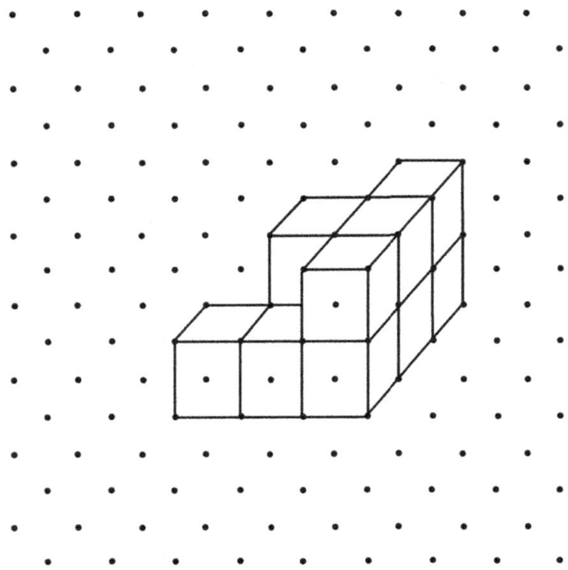

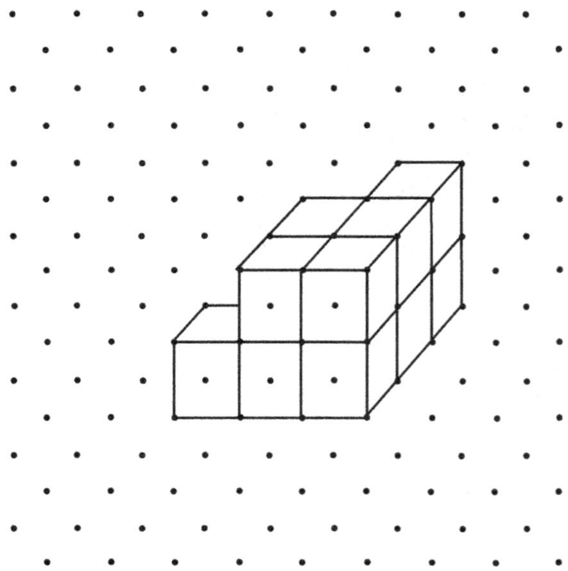

Marking Scheme Choose 1 of the drawings. 1 mark given if the provided Side View matches the chosen drawing. 1 mark given if the provided Front View matches the same drawing. 1 mark given if the provided Top View matches the same drawing. 1 mark if the second drawing matches all the three provided views. Total of 4 marks.

27.

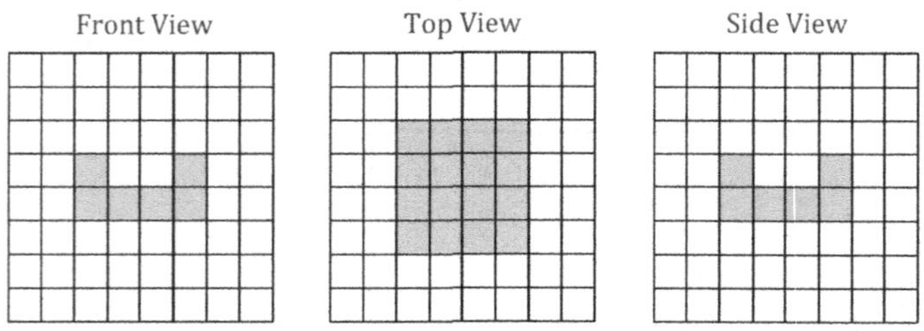

28. C

29a. v

29b. For this particular question, the orientation of the Top View does matter. Any other rotations of the shaded figure are incorrect. The exact placement of the shaded figure on the grid does not matter however.

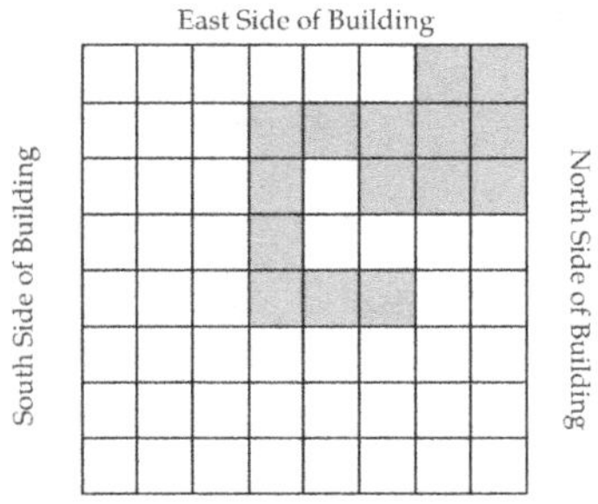

30.

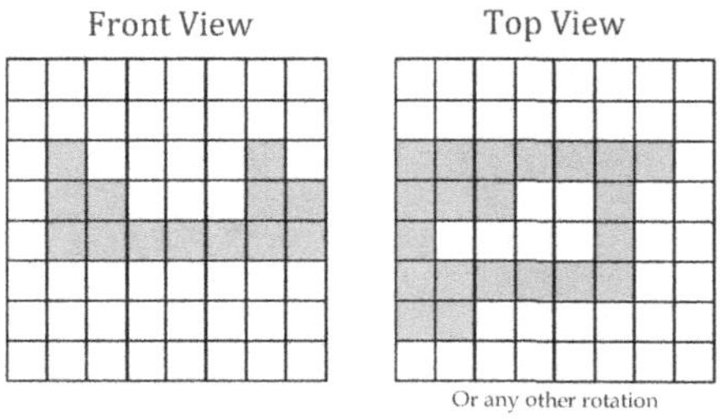

Check out this other Math is Key Workbook!

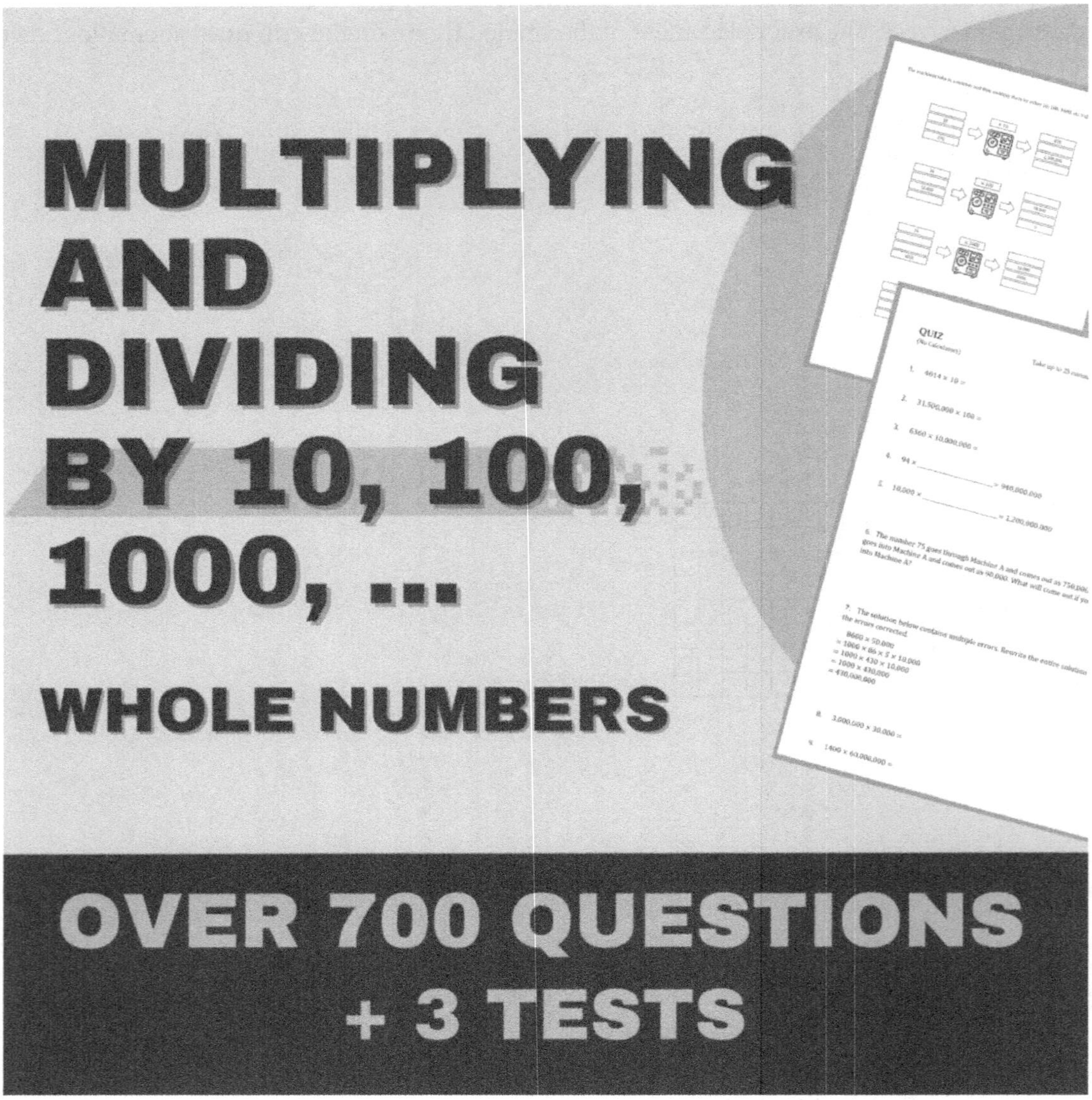

Visit www.MathisKey.com to access this workbook and other resources.

Made in United States
North Haven, CT
10 November 2026

22088875R00050